Energy Economics

Abdulkadir Develi / Selahattin Kaynak (eds.)

ENERGY ECONOMICS

PETER LANG
Frankfurt am Main · Berlin · Bern · Bruxelles · New York · Oxford · Wien

Bibliographic Information published by the Deutsche Nationalbibliothek
The Deutsche Nationalbibliothek lists this publication in the Deutsche Nationalbibliografie; detailed bibliographic data is available in the internet at http://dnb.d-nb.de.

Published with financial support
from Besel Industrial Products, Ankara

ISBN 978-3-631-63333-5

www.peterlang.de

Preface

Energy, being a very important part of human life, demonstrates its economic and political importance on a daily basis in countries' domestic and foreign policies. Energy plays an important role in shaping energy-based foreign policy in the West. The energy consumption of the EU, the world's largest energy consumer, continues to trend upwards. The last Ukrainian crisis forced the EU to search for new energy supplies. Alternative types of energy to replace Russia and the use of renewable energy resources are top issues of EU energy policies. Added to which, energy governance with limited resources is becoming increasingly important, both for the EU and the rest of the World. Turkey acts as a transit country amidst these energy struggles, and is in the process of negotiating to join the EU. This position encourages both Turkey-EU cooperation and further integration.

2012 Ankara / Turkey

Dr. Abdulkadir Develi
Dr. Selahattin Kaynak

Table of Contents

Increased Share of Natural Gas Consumption Compared with other Energy Resources; Structure of Natural Gas Market Demand and Supply

Selahattin Kaynak , İbrahim Arslan and A. Yilmaz Ata

1. Introduction

Development, growth and welfare are primary priorities for societies. Sustainable development, which is described as offering environmental, social and economic services to everyone without threatening the ecological end social systems, play a very important role with regard to these priorities. In this context, energy resources play an important role in sustainable development and growth.[1] Sustainable development is dependent on three basic concepts that are economic, social and environmental. Economic, social and environmental policies must be synchronized to achieve sustainable development.[2]

Energy is one of the fundamental aspects of economic, social and environmental policies.[3] The important role of energy comes from its structural relationship with all other sectors.[4] Because of population growth, technological improvements and rising standards of living, the need of energy has been growing and consequently, energy prices are increasing. The rise in the price of energy, which is a basic input for all areas of the economy, has negative effects on economic development and growth.[5] Therefore, to ensure sustainable development and economic growth, energy efficiency must be secured through stable and consistent energy policies. Besides, new and renewable energy resources must be utilized.[6] Sustainable development is interlinked with supply safety, and energy policies based on competition and protection of the environment. All of these factors must be harmonized and

---

1 Hall, C., et al. (2001), "The Need to Reintegrate the Natural Sciences with Economics", *Bioscience,* 51 (8), pp. 663-673.

2 Nordhaus, William D. (1973), "The Allocation of Energy Resources", *Brookings Papers on Economic Activity,* No. 3, pp. 529-579.

3 Goldember, J, et al., (2001), "Energy for the New Millennium", *Ambio*, 30 (6), pp. 330-337.

4 Toman, M. And B. Jemelkova (2003), "Energy and Development: An Assessment of the State of Knowledge", Resources for The Future Discussion Paper, 03-13.

5 Stern, D. I and Cleveland,-C.J (2004), "Energy and Economic Growth", *Rensselaer Working Papers in Economics* 0410, Rensselaer Polytechnic Institute, Department of Economics.

6 Imran, K and M. M. Siddiqui (2010), "Energy Consumption and Economic Growth: A Case Study of Three SAARC Countries", *European Journal of Social Sciences*, 16 (2), pp. 206-213.

national economic policies defined in consideration of global energy policies.[7] The main subject can be summarized as providing economic development, and ensuring a clean environment with productive and safe energy resources.

At the present, traditional energy resources are consumed rapidly. The possibility of an energy shortage in the future, and consequently higher energy prices, is prompting countries to find new energy resources. But new and alternative energy resources, which are seen as the solution to the energy bottleneck, account for only a very small share of current energy consumption. Besides, the necessary investments for energy resources such as hydrogen and fuel cells, and comprehensive problems, mean that traditional energy resources and fossil fuel are still the preferred energy resource.[8] For example according to the 2007-2030 projection of the International Energy Agency, even though the share of fossil fuel in total energy consumption will decrease to 80% from 81%, its weighting relative to total energy consumption will not change. In this context, energy resources must be used more efficiently and productively.

2. Efficient Use of Energy

The clean, cheap and safe provision of energy, an important parameter of economic development, is on the agenda of virtually every country.[9] With each passing day, international legislation governing environmental pollution is improving. R&D studies into cheap and clean energy continue. Political, commercial and military efforts must continue to strive to secure a sustainable energy supply.[10] Moreover, countries with poor energy resources are implementing policies on how to use current and renewable energy resources.[11]

A priority topic on most countries' agendas, sustainable energy is seen as an important factor for environmental, economic and development sustainability.[12] A sustainable energy approach comprises a continuous energy supply policy,

---

7 Truddea, N. and I. Murray, (2011), "Development of Energy Efficient Indicators in Russia", *IEA Working Paper*, International Energy Agency Press, p. 10.

8 Goldember, J, et al., (2001), "Energy for the New Millennium", *Ambio*, 30 (6), pp. 330-337.

9 (Nordhaus, William D. (1973), "The Allocation of Energy Resources", *Brookings Papers on Economic Activity,* No. 3, pp. 529-579.

10 Clawson, P. and S. Henderson, (2005), "Reducing Vulnerability to Middle East Energy Shocks", *Policy Focus*, No. 49, The Washington Institute for Near East Policy.

11 Goldember, J, et al., (2001), "Energy for the New Millennium", *Ambio*, 30 (6), pp. 330-337.

12 EIA (2011), Monthly Energy Review: January 2011, U.S. Energy Information Administration (EIA) Press.

technology and applications. Sustainability in terms of energy relies on three main principles:

1. Efficient use of energy and energy saving.
2. Development of environmentally friendly strategies to minimize the harmful effects of energy consumption.
3. Promoting renewable energy resources and improving technological ability in this field.

Energy productivity and saving must be pursued at all levels, from production to consumption. Today, energy intensity is seen as an indicator of development in many countries.[13] Low energy intensity indicates productivity and energy efficiency. Thus, the cost of energy will decrease, and economic growth and development will subsequently be affected positively.[14] At the same time, progress in energy productivity will reduce energy consumption and the demand of energy will decline. Therefore, it will have a positive effect on resolving the energy bottleneck.[15] It is very important for a country to provide continuous and low-cost energy in order to achieve sustainable development.[16] Currently, numerous countries are preparing national programs for achieving sustainable development through sustainable energy. Liberal models are aiming to provide such energy policies, which are struggling to cope with the economic boom and energy dependence.[17]

3. The Causes of the Demand for Energy

It is predicted that world energy demand will increase as a result of economic development and population growth. It is also predicted that an important share of energy demand will stem from developing countries.[18]

As shown in Table 1 and Figure 1, according to the forecast issued by the International Energy Agency in 2007, 12 billion tons of oil-equivalent energy consumption will increase 40% and reach 16.8 billion tons in 2030. The annual

13 Truddea, N. and I. Murray, (2011), "Development of Energy Efficient Indicators in Russia", *IEA Working Paper*, International Energy Agency Press.

14 Stern, David I. (1999), "Is Energy Cost an Accurate Indicator of Natural Resource Quality?" *Ecological Economics*, 31, pp. 381-394.

15 Stern, David I. (1993), "Energy Use and Economic Growth in the USA, Multivariate Approach", *Energy Economics*, 15, pp. 137-150.

16 Goldember, J, et al., (2001), "Energy for the New Millennium", *Ambio*, 30 (6), pp. 330-337.

17 Ausvik, Ole G. (2009), "EU Natural Gas Market Liberalization and Long-Term Security of Supply and Demand", *Political Economy of Energy in Europe* (Ed.) G. Femann, pp. 85-115.

18 Goldember, J, et al., (2001), "Energy for the New Millennium", *Ambio*, 30 (6), pp. 330-337.

average rate of increase is calculated as 1.5% over the same period. Over the projection period, oil consumption is predicted to increase at a rate of 0.9%, and natural gas at a rate of 1.5%.[19]

Table 1: Change in Primary Energy Demand in the Reference Scenario (Mtoe)

	1980	2000	2007	2015	2030	2007-2030(%)
Coal	1792	2292	3184	3828	4887	1.9
Oil	3107	3655	4093	4234	5009	0.9
Natural Gas	1234	2085	2512	2801	3561	1.5
Nuclear	186	676	709	810	956	1.3
Hydro	148	225	265	317	402	1.8
Biomass	749	1031	1176	1338	1604	1.4
Other	12	55	74	160	370	7.3
TOTAL	7228	10019	12013	13488	16789	1.5

Reference: IEA WEA 2009, Reference Scenario

Fig. 1: Share of Primary Energy Resources in Energy Consumption (2007-2030)

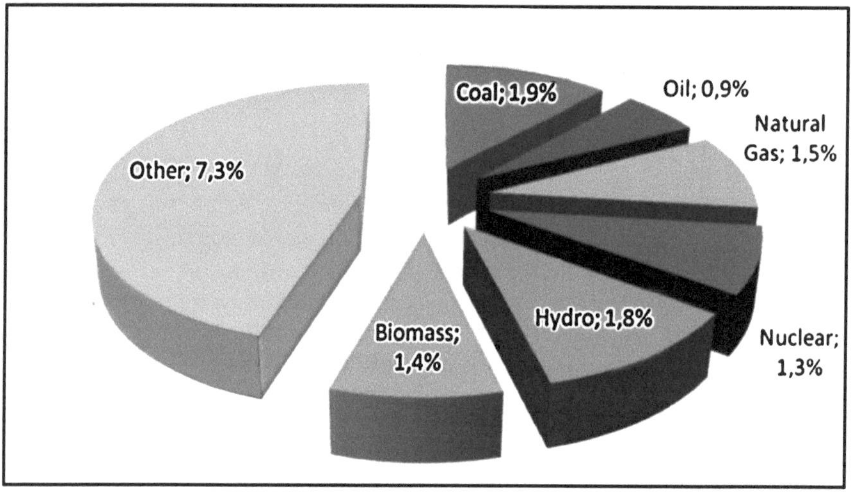

Reference: IEA WEA 2009, Reference Scenario.

With urbanization and population growth, needs have increased and consumption has risen. The industry has improved rapidly as a result of the diversified

19 IEA (2009), World Energy Outlook: Executive Summary, International Energy Agency Press.

needs. In addition to rapid industrial improvements, technological developments have resulted in production machinery with very high capacity levels. With the introduction of these machines, the need for energy has been increasing.[20] Most of the energy is obtained from coal, oil, natural gas, etc. (known as fossil fuel).[21] Fossil fuel plays a key role compared with all other energy resources. These conventional energy resources are used in all fields of modern life. Over the last two decades, fossil fuel took priority over renewable energy resources due to its cheaper price and more mature production technologies. Recently, interest in natural gas has increased.[22] As shown in Figure 2, which is a projection for 2030, the share of fossil fuels (coal, oil, natural gas, nuclear, hydro, biomass and others) in total energy consumption will remain the same. Oil (5009 billion m^3), coal (4887 billion m^3) and natural gas (3561 billion m^3) will be the most preferable among all kinds of fuels.

Fig. 2: Share of primary energy resources in energy consumption (2030; Mtoe)

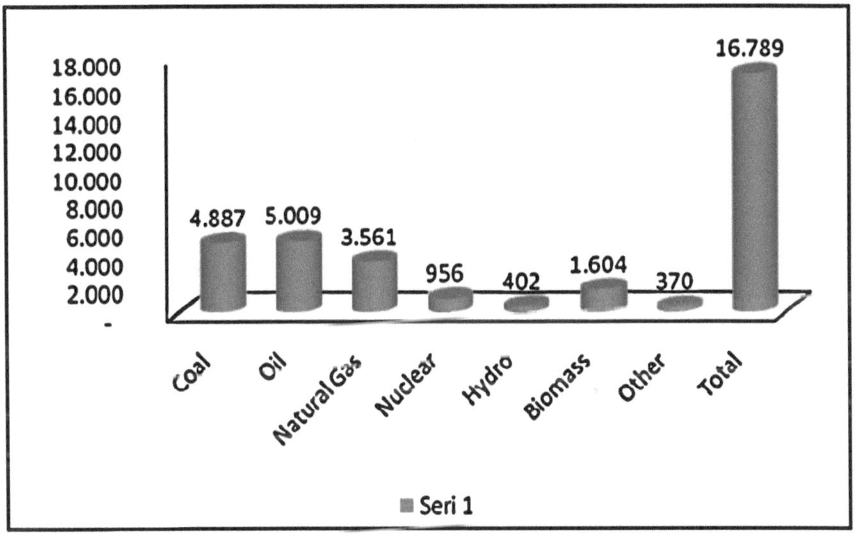

Reference: IEA WEA 2009, Reference Scenario.

20 Rees R. and S. Scholz (2010), "Electricity Market Design for Germany", SosyoEkonomi, 6 (12), Enerji Özel Sayisi, pp. 39-62.

21 Goldember, J, et al., (2001), "Energy for the New Millennium", *Ambio*, 30 (6), pp. 331.

22 Nordhaus, William D. (1973), "The Allocation of Energy Resources", *Brookings Papers on Economic Activity,* No. 3, pp. 533:

4. The Importance of Natural Gas and the Improvement Process

Of all fossil fuels, oil, natural gas and coal are the most important energy sources. Because of their advantages, their share and importance compared with other energy resources have been increasing. The need of energy is an important factor in rapid and sustainable economic development.[23] It is predicted that up to 2030 global energy demand will increase 1.8%, while demand for natural gas will increase 2.1%.[24] As a raw material in some chemical products, natural gas accounts for a large share of energy consumption in the world. The history of natural gas goes back hundreds of years. Historical sources show that natural gas was first used around 900 BC in China. Natural gas – which is easy to carry, process and stock – began to be used commonly in England in 1790. With the advent of the Industrial Revolution, natural gas became a material of strategic and economic importance. As pipelines were built, the consumption of natural gas increased in the 1920s and became even more prevalent after the Second World War.[25] In light of both technological improvements and environmental concerns, natural gas surpassed the usage of oil. Natural gas was first used in the USA in the energy sector. But its importance as an energy resource became clear just after the energy crisis in 1970.[26] Since then, economic and technological developments have prompted heightened interest in new and renewable energy resources. At the beginning of the 1970s, in particular, the Organization of Petroleum Exporting Countries raised the prices of oil rapidly, resulting in the Petroleum Crisis and, finally, strengthened focus on new energy resources. Back then, political military conflicts between Arab countries and Israel forced the Arab countries to ally with each other. The OPEC union used its petroleum reserves as a political weapon against Israel and other countries that supported Israel. Their intention was to gain the support of oil-dependent industrialized western countries by raising the price of crude oil. But the policies implemented by OPEC prompted Western Europe and North America to expend more effort on inventing and using new energy resources. Moreover, many developed countries – irrespective of the energy bottleneck – have realized the disadvantages of energy shortages and the danger of being energy dependent. This realization has resulted in developed western

23 MIT (2010); "The Future of Natural Gas", *Interim Report*, Massachusetts Institute of Technology (MIT) Pres, p. 1.

24 Kavalov B., H. Petric and A. Georgakaki (2009); "Liquefied Natural Gas for Europe-Some Important Issues for Consideration", *JRC References Reports*, European Commission (JRC). p. 9.

25 MIT (2010); "The Future of Natural Gas", *Interim Report*, Massachusetts Institute of Technology (MIT) Pres,p.68

26 Ausvik, Ole G. (2009), "EU Natural Gas Market Liberalization and Long-Term Security of Supply and Demand", *Political Economy of Energy in Europe* (Ed.) G. Femann, pp. 86-87.

countries allocating bigger budgets and more labour to finding new energy resources.[27] One of the newest energy alternatives was natural gas. The share of natural gas in the total world energy sector has increased after the two petroleum crisis that occurred concurrently with the Yom Kippur War in 1973 and the Iranian Revolution in 1979. While the share of natural gas in total world energy consumption used to be less than 10%, it nowadays accounts for 24% percent of total energy consumption. The reason why natural gas is more preferable than oil is that it is less harmful to nature. Moreover, compared to oil, natural gas is easier to produce and transport than oil. The known natural gas reserves will last for almost 70 years. The known oil reserves are equivalent to petroleum reserves.[28]

5. Natural Gas Market

5.1 Structure of supply

With every day that passes, energy proves how indispensable it is for development. Today, improved energy efficiency is a must. Interconnecting countries that supply energy with countries that demand it has become essential. An analysis of energy maps reveals a grid of pipelines that cover the countries like a network. This circumstance implies that not only the supply of natural gas but also other energy resources are sophisticated.[29]

The current natural gas reserves are not distributed evenly around the world. Due to the location of the reserves, the production of natural gas is intensified in some regions and countries. This intensity creates monopolistic structures on the supply side of natural gas. In recent years, since the supply of energy has been limited to some regions and countries, controlling the price of natural gas by monopolistic behaviours has caused the problem of energy supply security.[30] The liberalization of natural gas markets is still in progress. The privatization of natural gas supply, which was formerly provided by the state, is a point at issue now. The public sector does not deal with energy supply services in many countries.[31]

27 Özsabuncuoğlu İ. H. and A. Uğur (2005), Doğal Kaynaklar: ekonomi, Yönetim ve Politika, İmaj Yayinlari, Ankara, pp. 157-158.

28 Clawson, P. and S. Henderson, (2005), "Reducing Vulnerability to Middle East Energy Shocks", *Policy Focus*, No. 49, The Washington Institute for Near East Policy, pp. 32.

29 MIT (2010), "The Future of Natural Gas", *Interim Report*, Massachusetts Institute of Technology (MIT) Press. p. 5.

30 MIT (2010), "The Future of Natural Gas", *Interim Report*, Massachusetts Institute of Technology (MIT) Press. p. 31.

31 Goldember et al., 2001, 333.

As multinational companies, especially Shell and BP, have entered the natural gas market, the market structure has become oligopolistic. This will cause different supply prices in different regions.[32]

5.2 Demand structure

Natural gas is preferred to other energy resources because it is less harmful to the environment than the others. This increases the demand for natural gas.[33] Natural gas is an important type of energy resource used in accommodation, trade and industry.[34] Therefore, natural gas is crucially important as an energy source. Due to this structure, the price elasticity of natural gas is low and income elasticity is limited. When the price of natural gas increases, consumers can't restrict their consumption and an important part of their budget goes on natural gas. The residential sector requires natural gas for heating. In addition, the consumption of electricity indirectly affects demand for natural gas demand. Since natural gas is used as a raw material, it is difficult to substitute in industry. In recent years, the consumption of natural gas for the production of electricity has triggered an increase in demand of natural gas during the summer months as well.[35] Various factors affecting the demand of natural gas, such as:

Demand fluctuates seasonally due to weather conditions, increasing in cold months, and decreasing in hot months. This fluctuation in demand is due to the fact that the residential and trade sectors utilize natural gas for heating. This means that demand of natural gas is highest in winter and lowest in summer. However, the increasing use of natural gas in the production of electricity is moderating the fluctuations in demand. In hot summer months, the demand of natural gas for heating decreases, but as the need for cooling rises, which increases the demand for electricity, so demand of natural gas also rises.[36]

Economic fluctuations in the industry sector play an important role in influencing the amount of natural gas that is consumed. When an economy is expanding, industrial output also increases. In recessions, industrial output also

---

32 MIT (2010), "The Future of Natural Gas", *Interim Report*, Massachusetts Institute of Technology (MIT) Press. p. 30.

33 Hanore A. (2006), "Future Natural Gas Demand in Europe: The Importance of the Power Sector", *Natural Gas Research Programme*, No. 1/103, Oxford Institute for Energy Studies. p. 28.

34 MIT (2010), "The Future of Natural Gas", *Interim Report*, Massachusetts Institute of Technology (MIT) Press. p. 39.

35 MIT (2010), "The Future of Natural Gas", *Interim Report*, Massachusetts Institute of Technology (MIT) Press. p. 53.

36 Hanore A. (2006); "Future Natural Gas Demand in Europe: The Importance of the Power Sector", *Natural Gas Research Programme*, No. 1/103, Oxford Institute for Energy Studies. p. 28.

decreases. The fluctuations in industrial production – depending on economic expansion or recession – affect demand of natural gas.[37]

Population growth not only directly affects the demand for, and composition of, energy but also indirectly affects economic growth and development. From a microeconomic point of view, population growth affects the demand by households for natural gas directly, while indirectly affecting demand for natural gas in production.[38]

Governments' energy policies and other policies influence energy demand significantly. In recent years, governments' environment policies have generalized the use of natural gas. Governments have put some policies in place to make the sector more competitive, which has resulted in an increased demand of energy. Moreover, government taxes on energy impact the consumption of energy resources through prices.[39]

6. World Natural Gas Reserves, Production, and Distribution

Natural gas reserves are abundant in the world.[40] But these reserves are not distributed evenly. Natural gas production is intensified in some regions and countries.[41]

Tables 2 and 3 show the distribution of natural gas reserves throughout countries and regions in 1989, 1999, 2008 and 2009. According to the table, Russia is home to ¼ of the world's natural gas reserves. Of the total natural gas reserves in the world, ¾ are in the Middle East, including Qatar, and in Europe & Eurasia, including Russia. Iran and Qatar hold 73% of the Middle East's natural gas reserves. England and Holland have 60% of the total EU reserves. Norway, which is not a member of the EU, has more reserves than the EU.

---

37 Hanore A. (2006); "Future Natural Gas Demand in Europe: The Importance of the Power Sector", *Natural Gas Research Programme*, No. 1/103, Oxford Institute for Energy Studies. p. 29.

38 MIT (2010); "The Future of Natural Gas", *Interim Report*, Massachusetts Institute of Technology (MIT) Press. p. 55.

39 (Hanore A. (2006); "Future Natural Gas Demand in Europe: The Importance of the Power Sector", *Natural Gas Research Programme*, No. 1/103, Oxford Institute for Energy Studies. p. 30.

40 MIT (2010); "The Future of Natural Gas", *Interim Report*, Massachusetts Institute of Technology (MIT) Press. p. 7.

41 BP (2010); BP Statistical Review of World Energy 2010. p. 22.

Table 2: Proved Reserves (Natural Gas)

	At end 1989	At end 1999	At end 2008	At end 2009			
	Trillio cubic metres	Trillion cubic metres	Trillion cubic metres	Trillion cubic feet	Trillion cubic metres	Share of total	R/P ratio
US	4.73	4.74	6.93	244.7	6.93	3.7%	11.7
Canada	2.73	1.72	1.75	62.0	1.75	0.9%	10.9
Mexico	2.06	0.86	0.50	16.8	0.48	0.3%	8.2
Total North America	9.52	7.32	9.18	323.4	9.16	4.9%	11.3
Argentina	0.67	0.73	0.40	13.2	0.37	0.2%	9.1
Bolivia	0.12	0.52	0.71	25.1	0.71	0.4%	57.9
Brazil	0.11	0.23	0.36	12.7	0.36	0.2%	30.4
Colombia	0.11	0.19	0.11	4.4	0.12	0.1%	11.8
Peru	0.34	0.25	0.25	11.2	0.32	0.2%	91.3
Trinidad & Tobago	0.29	0.61	0.44	15.4	0.44	0.2%	10.7
Venezuela	2.99	4.15	4.98	200.1	5.67	3.0%	*
Other S. & Cent. America	0.17	0.14	0.07	2.5	0.07	◆	19.7
Total S. & Cent. America	4.80	6.81	7.32	284.6	8.06	4.3%	53.2
Azerbaijan	n/a	1.23	1.31	46.3	1.31	0.7%	88.8
Denmark	0.11	0.14	0.06	2.3	0.06	◆	7.6
Germany	0.29	0.23	0.10	2.7	0.08	◆	6.4
Italy	0.30	0.19	0.06	2.3	0.06	◆	8.6
Kazakhstan	n/a	1.78	1.82	64.4	1.82	1.0%	56.6
Netherlands	1.57	1.55	1.14	38.3	1.09	0.6%	17.3
Norway	1.73	1.25	2.22	72.3	2.05	1.1%	19.8
Poland	0.16	0.15	0.11	3.8	0.11	0.1%	26.6
Romania	0.13	0.35	0.63	22.2	0.63	0.3%	57.9
Russian Federation	n/a	42.44	43.30	1567.1	44.38	23.7%	84.1
Turkmenistan	n/a	2.59	8.10	286.2	8.10	4.3%	*
Ukraine	n/a	1.01	0.98	34.7	0.98	0.5%	51.0
United Kingdom	0.56	1.27	0.29	10.3	0.29	0.2%	4.9
Uzbekistan	n/a	1.58	1.68	59.4	1.68	0.9%	26.1
Other Europe & Eurasia	47.41	0.43	0.45	15.7	0.44	0.2%	46.6
Total Europe & Eurasia	52.28	56.17	62.26	2228.1	63.09	33.7%	64.8
Bahrain	0.18	0.12	0.09	3.0	0.09	◆	6.7
Iran	17.00	25.00	29.61	1045.7	29.61	15.8%	*
Iraq	3.12	3.29	3.17	111.9	3.17	1.7%	*
Kuwait	1.40	1.48	1.78	63.0	1.78	1.0%	*

	At end 1989	At end 1999	At end 2008	At end 2009			
	Trillio cubic metres	Trillion cubic metres	Trillion cubic metres	Trillion cubic feet	Trillion cubic metres	Share of total	R/P ratio
Oman	0.28	0.83	0.98	34.6	0.98	0.5%	39.6
Qatar	4.62	11.16	25.37	895.8	25.37	13.5%	*
Saudi Arabia	5.22	6.15	7.57	279.7	7.92	4.2%	*
Syria	0.16	0.24	0.28	10.0	0.28	0.2%	48.9
United Arab Emirates	5.65	6.00	6.43	227.1	6.43	3.4%	*
Yemen	0.20	0.48	0.49	17.3	0.49	0.3%	*
Other Middle East	0.01	0.01	0.05	2.2	0.06	◆	24.2
Total Middle East	37.83	54.74	75.82	2690.4	76.18	40.6%	*
Algeria	3.25	4.52	4.50	159.1	4.50	2.4%	55.3
Egypt	0.35	1.22	2.15	77.3	2.19	1.2%	34.9
Libya	1.22	1.32	1.54	54.4	1.54	0.8%	*
Nigeria	2.83	3.51	5.25	185.4	5.25	2.8%	*
Other Africa	0.82	0.87	1.27	45.0	1.27	0.7%	65.3
Total Africa	8.48	11.44	14.71	521.2	14.76	7.9%	72.4
Australia	0.96	1.99	3.08	108.7	3.08	1.6%	72.7
Bangladesh	0.73	0.32	0.34	12.5	0.36	0.2%	18.0
Brunei	0.32	0.37	0.35	12.4	0.35	0.2%	30.7
China	1.02	1.37	2.46	86.7	2.46	1.3%	28.8
India	0.69	0.65	1.09	39.4	1.12	0.6%	28.4
Indonesia	2.55	2.62	3.18	112.5	3.18	1.7%	44.3
Malaysia	1.61	2.48	2.38	84.1	2.38	1.3%	38.0
Myanmar	0.27	0.29	0.57	20.1	0.57	0.3%	49.4
Pakistan	0.65	0.70	0.84	32.0	0.91	0.5%	23.9
Papua New Guinea	0.23	0.43	0.44	15.6	0.44	0.2%	*
Thailand	0.23	0.35	0.34	12.7	0.36	0.2%	11.6
Vietnam	–	0.17	0.56	24.1	0.68	0.4%	85.2
Other Asia Pacific	0.26	0.34	0.36	12.9	0.36	0.2%	20.9
Total Asia Pacific	9.50	12.07	16.00	573.6	16.24	8.7%	37.0
Total World	122.40	148.55	185.28	6621.2	187.49	100.0%	62.8
Of which: EuropeanUnion	3.36	3.98	2.49	85.5	2.42	1.3%	14.1
OECD	15.60	14.26	16.44	571.2	16.18	8.6%	14.4
FormerSovietUnion	47.06	50.85	57.46	2067.1	58.53	31.2%	84.2

Reference: BP Statistical Review of World Energy 2010, pp. 22

Table 3: Distribution of Proved Reserves by Regions

	At end 1989 (Tcm)*	At end1999 (Tcm)	At end 2008 (Tcm)	At end 2009 (Tcm)
Total North America	9.52	7.32	9.18	9.16
Total S. & Cent. America	4.80	6.81	7.32	8.06
Total Europe & Eurasia	52.28	56.17	62.26	63.09
Total Middle East	37.83	54.74	75.82	76.18
Total Africa	8.48	11.44	14.71	14.76
Total Asia Pacific	9.50	12.07	16.00	16.24
Total World	122.40	148.55	185.28	187.49

* *Trillion cubic metres*

Reference: *BP Statistical Review of World Energy 2010, pp. 22*

The world's natural gas reserves by percentage weight in the regions are shown in Figure 3. Accordingly, the Middle East has the largest share of the world's natural gas reserves, with 40%, followed by the European region & Russia with a share of 34%.[42]

Fig. 3: Distribution of Proved Reserves by Regions (%)

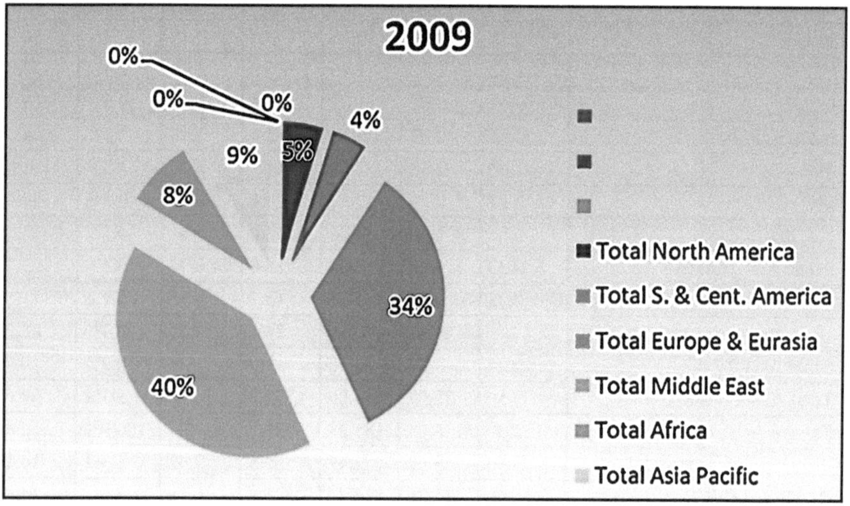

Reference: BP Statistical Review of World Energy 2010, pp. 22

42 BP (2010); BP Statistical Review of World Energy 2010. p. 22

The natural gas reserves and amount produced in 2008 are shown in Table 4 and Figure 4. In 2008, North America was home to 4.8% of the world's natural gas reserves, with 8.9 trillion cubic metres, but accounted for 26.5 percent of world production, with 812.3 trillion cubic metres. As such, North America has a very high production-reserve ratio. The opposite is true of the Middle East, which is home to 41 percent of the world's reserves but accounts for only 12.4 percent of world natural gas production, with 381.1 trillion cubic metres. This indicates that the Middle East's importance in natural gas production will increase. Nevertheless, Europe & Eurasia, which hold 34% of the world's natural gas reserves amounting to 62.9 trillion cubic metres, and account for 35.5% of world natural gas production, with 1087.3 trillion cubic metres in total, are the most important players in the global natural gas market. The EU has only 1.6 percent of the world's natural gas reserves, and accounts for 6.2 percent of world natural gas production.[43]

Table 4: Natural Gas Reserves and Amount of Production (2008)

	Proved Reserves (trillion $m^{3)}$	Production (trillion $m^{3)}$
Total Middle East	75.9	381.1
Europe & Eurasia	62.9	1.087.3
Total Africa	14.7	214.8
Total Asia-Pacific	15.4	411.2
Total North America	8.9	812.3
Total South & Cent. America	7.3	158.9
EU	2.9	190.3
Total World	185	3.065.6

Reference: BP Statistical Review of World Energy 2009

43 BP (2010), BP Statistical Review of World Energy 2010.

Fig. 4: Distribution of Proved Reserves and Amount of Production by Regions (2008)

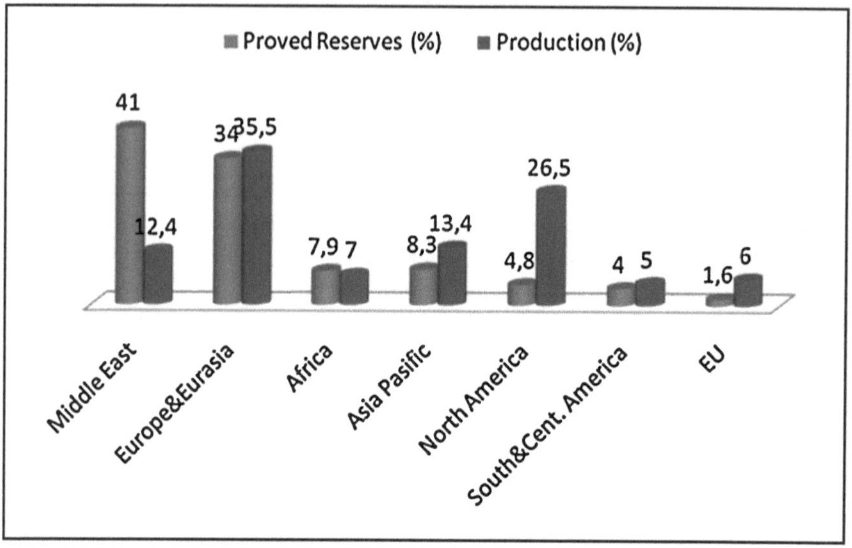

Reference: BP Statistical Review of World Energy 2009

According to data from the International Energy Agency, natural gas production is expected to decline in the period 2007-2030, although it is expected to increase in other regions outside Europe. After 2020, the increase in production in North America is expected to slow down. The Middle East and Africa are regions where large increases in production are expected during the same projection period.[44]

7. World Natural Gas Consumption and Distribution

Natural gas consumption has been increasing rapidly because of the fact that natural gas emits less pollution than other fossil fuels.[45] World natural gas consumption increased 3.5% between 1965 and 2007. When regional rises are analyzed in the same period, the ratios are 12% in North America, 5.6% in South

44 IEA (2009); World Energy Outlook: Executive Summary, International Energy Agency Press.

45 Clawson, P. and S. Henderson, (2005); "Reducing Vulnerability to Middle East Energy Shocks", *Policy Focus*, No. 49, The Washington Institute for Near East Policy, p.32.

and Middle America, 5.2% in Europe and Eurasia, 8.7% in Middle East, 12.2% in Africa, and 11.5% in South of Asia-Pacific. World natural gas consumption increased 3.3% in 2007. As the world's largest market, the USA stagnated because of high prices and industrial restructuring conditions. In some countries, such as Russia, China and the Middle East, natural gas consumption increased 4%. Until 1980, most natural gas was consumed in North America, but Europe and Eurasia have been consuming most since 1980.[46] Table 5 shows the regional distribution of natural gas consumption over the years. Analyzing the data in the table shows that, between 1999 and 2009, natural gas consumption increased not only in the regions, but also in the world. For example, global natural gas consumption was 2321.3 million cubic metres in 1999 compared with 2940.4 million cubic metres in 2009.

46 IEA (2009); World Energy Outlook: Executive Summary, International Energy Agency Press.

Table 5: Consumption of Natural Gas by Regions (1999-2009; billion m^3)

Billion cubic metres	1999	2000	2001	2002	2003	2004	2005	2006	2007	2008	2009
Total North America	759.2	793.7	758.7	787.4	779.1	782.4	774.7	771.9	813.9	822.0	810.9
Total S. & Cent. America	89.5	96.0	100.7	102.1	107.9	117.5	122.5	135.3	138.1	141.0	134.7
Total Europe & Eurasia	967.8	984.6	1014.5	1018.3	1060.5	1089.4	1114.2	1121.4	1135.5	1138.5	1058.6
Total Middle East	181.1	186.7	206.8	217.6	229.0	247.1	279.2	291.5	303.1	331.8	345.6
Total Africa	53.4	57.2	62.6	64.7	71.5	77.4	79.4	84.1	90.8	96.1	94.0
Total Asia Pacific	270.3	292.6	312.1	326.4	352.8	367.6	397.5	425.3	455.8	481.4	496.6
Total World	2321.3	2410.8	2455.3	2516.5	2600.9	2681.4	2767.5	2829.5	2937.1	3010.8	2940.4

Reference: BP Statistical Review of World Energy 2010, pp. 27

Figure 5 shows the regional distribution of natural gas consumption. Europe and Eurasia consumed the most, at 1058.6 billion cubic metres, followed by North America, with 810.9 billion cubic metres.[47]

Fig. 5: Distribution of Consumption by Regions (2009;%)

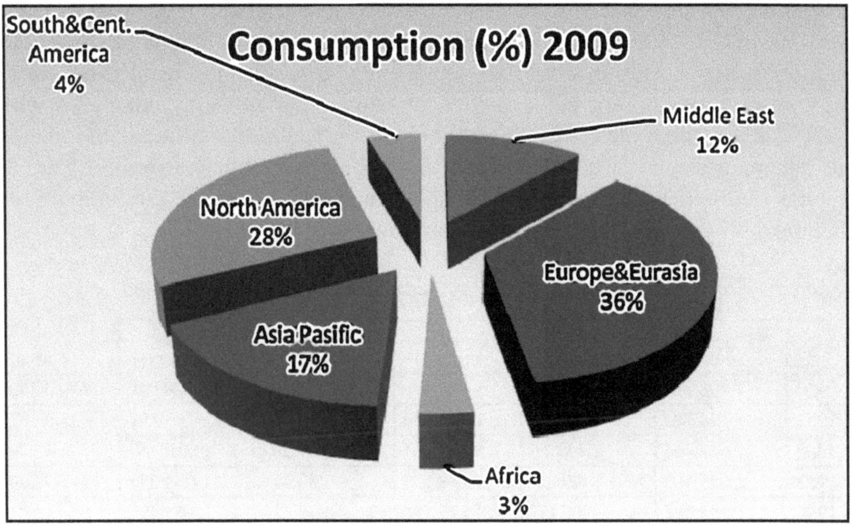

Reference: BP Statistical Review of World Energy 2010, p. 27

It is predicted that demand of natural gas will increase in Africa, Latin America and developing Asia. It is expected that natural gas demand in Latin America will increase 4.4% and reach 335 billion cubic metres in 2030, following 102 billion cubic metres in 2002. Natural gas demand is expected to increase rapidly between 2002 and 2030 at an average rate of 5.1% annually. Demand in developing Asia is predicted to increase faster than in other regions, and to reach 672 billion cubic metres.[48]

47 BP (2010), BP Statistical Review of World Energy 2010.

48 IEA (2010); Energy Efficiency Governance, Handbook Second Edition, International Energy Agency Press.

8. World Natural Gas Consumption by Sectors

Natural gas is an important source of energy that is used in the electricity, residential, industrial and transportation sectors, and consumption in these sectors is increasing.[49] Table 6 shows the total consumption of natural gas and the amount consumed by the electricity, residential, trade, industrial and transportation sectors between 1950 and 2009. The data in the table indicates that natural gas consumption has increased across all sectors. For example, the residential sector consumed 1.198 billion cubic metres of natural gas in 1950, compared with 4.760 billion cubic metres in 2009. Similarly, in 1950, 628 billion cubic metres of natural gas were used in power plants, compared with 6.887 billion cubic metres in 2009. What is remarkable about the table is that the consumption of natural gas by power plants has increased most.[50]

Table 6: Natural Gas Consumption by Sector, (1949-2009; billion m3)

	Residential Sector	Commercial Sector	Industrial Sector	Transportation Sector	Electric Power Sector	Total
1950	1.198.369	387.838	3.425.870	125.546	628.919	5.766.542
1951	1.474.725	464.309	3.914.734	192.496	763.898	6.810.162
1952	1.621.966	515.669	4.039.361	207.207	910.117	7.294.320
1953	1.685.503	530.65	4.158.531	230.314	1.034.272	7.639.270
1954	1.894.248	584.957	4.173.186	230.615	1.165.498	8.048.504
1955	2.123.952	629.219	4.541.960	245.246	1.153.280	8.693.657
1956	2.327.564	716.871	4.709.147	295.972	1.239.311	9.288.865
1957	2.500.269	775.916	4.934.578	299.235	1.336.141	9.846.139
1958	2.714.251	871.774	5.031.509	312.221	1.372.853	10.302.608
1959	2.912.601	975.107	5.455.616	349.348	1.628.509	11.321.181
1960	3.103.167	1.020.222	5.771.311	347.075	1.724.762	11.966.537
1961	3.248.578	1.076.849	5.961.117	377.607	1.825.117	12.489.268
1962	3.478.563	1.206.668	6.232.812	382.496	1.965.974	13.266.513
1963	3.589.021	1.267.783	6.545.169	423.783	2.144.473	13.970.229
1964	3.787.292	1.374.717	6.893.333	435.57	2.322.896	14.813.808
1965	3.902.802	1.443.648	7.111.641	500.524	2.321.101	15.279.716
1966	4.138.259	1.622.740	7.546.102	535.353	2.609.949	16.452.403
1967	4.313.304	1.958.970	7.793.982	575.752	2.746.352	17.388.360

49 MIT (2010); "The Future of Natural Gas", *Interim Report*, Massachusetts Institute of Technology (MIT) Press.-Nordhaus, William D. (1973); "The Allocation of Energy Resources", *Brookings Papers on Economic Activity*, No. 3, pp. 539.

50 IEA (2011); Energy Efficiency Governance, Handbook Second Edition, International Energy Agency Press

	Residential Sector	Commercial Sector	Industrial Sector	Transportation Sector	Electric Power Sector	Total
1968	4.450.354	2.075.736	8.367.098	590.965	3.147.909	18.632.062
1969	4.728.281	2.253.206	8.956.149	630.962	3.487.642	20.056.240
1970	4.837.432	2.398.510	9.249.418	722.166	3.931.860	21.139.386
1971	4.971.690	2.508.977	9.594.177	742.592	3.976.018	21.793.454
1972	5.125.982	2.607.982	9.624.418	766.156	3.976.913	22.101.451
1973	4.879.387	2.597.037	10.184.590	728.177	3.660.172	22.049.363
1974	4.786.128	2.555.617	9.769.168	668.792	3.443.428	21.223.133
1975	4.924.124	2.508.293	8.364.544	582.963	3.157.669	19.537.593
1976	5.051.360	2.667.740	8.598.205	548.323	3.080.868	19.946.496
1977	4.821.485	2.500.793	8.474.434	532.669	3.191.200	19.520.581
1978	4.903.006	2.601.106	8.404.552	530.451	3.188.363	19.627.478
1979	4.965.365	2.785.961	8.397.948	600.964	3.490.523	20.240.761
1980	4.752.082	2.610.895	8.198.099	634.622	3.681.595	19.877.293
1981	4.546.450	2.519.791	8.055.138	642.325	3.640.154	19.403.858
1982	4.633.035	2.605.523	6.940.568	596.411	3.225.518	18.001.055
1983	4.380.599	2.432.547	6.620.957	490.042	2.910.767	16.834.912
1984	4.555.465	2.524.244	7.230.722	528.754	3.111.342	17.950.527
1985	4.433.377	2.432.382	6.867.335	503.766	3.044.083	17.280.943
1986	4.313.969	2.318.335	6.501.581	485.041	2.602.370	16.221.296
1987	4.314.833	2.430.064	7.102.691	519.17	2.844.051	17.210.809
1988	4.630.330	2.670.465	7.479.265	613.912	2.635.613	18.029.585
1989	4.780.638	2.717.722	7.886.146	629.308	3.105.183	19.118.997
1990	4.391.324	2.622.721	8.254.806	660.086	3.244.619	19.173.556
1991	4.555.659	2.728.581	8.360.230	601.672	3.315.925	19.562.067
1992	4.690.065	2.802.751	8.697.719	589.822	3.447.871	20.228.228
1993	4.956.445	2.861.569	8.871.678	627.168	3.472.982	20.789.842
1994	4.847.701	2.895.013	8.913.269	688.569	3.902.546	21.247.098
1995	4.850.318	3.031.077	9.384.048	704.92	4.236.526	22.206.889
1996	5.241.414	3.158.244	9.685.008	717.513	3.806.901	22.609.080
1997	4.983.772	3.214.912	9.714.058	759.798	4.064.803	22.737.343
1998	4.520.276	2.999.491	9.493.087	644.818	4.588.284	22.245.956
1999	4.725.672	3.044.658	9.158.348	656.941	4.819.531	22.405.150
2000	4.996.179	3.182.469	9.293.188	654.962	5.206.324	23.333.122
2001	4.771.340	3.022.712	8.462.771	639.5	5.342.301	22.238.624
2002	4.888.818	3.144.170	8.620.262	681.87	5.671.897	23.007.017
2003	5.079.351	3.179.493	8.272.679	609.763	5.135.215	22.276.501
2004	4.868.797	3.128.972	8.340.741	586.701	5.463.763	22.388.974
2005	4.826.775	2.998.920	7.708.847	606.91	5.869.145	22.010.597
2006	4.368.466	2.832.030	7.654.092	607.952	6.222.100	21.684.641
2007	4.722.358	3.012.904	7.874.449	646.02	6.841.408	23.097.140

	Residential Sector	Commercial Sector	Industrial Sector	Transportation Sector	Electric Power Sector	Total
2008	4.872.107	3.135.852	7.874.062	676.213	6.668.379	23.226.612
2009	4.760.647	3.113.007	7.403.654	668.868	6.887.907	22.834.083

Reference: EIA/Monthly Energy Review, January 2011

As of 2009, Figure 6 shows the distribution of natural gas consumption on a sector-specific basis. Accordingly, industry accounts for the highest consumption, at 32%, followed by electricity, at 30%, and the residential sector, at 21%.[51]

Fig. 6: Distribution of Natural Gas Consumption by Sector (2009;%)

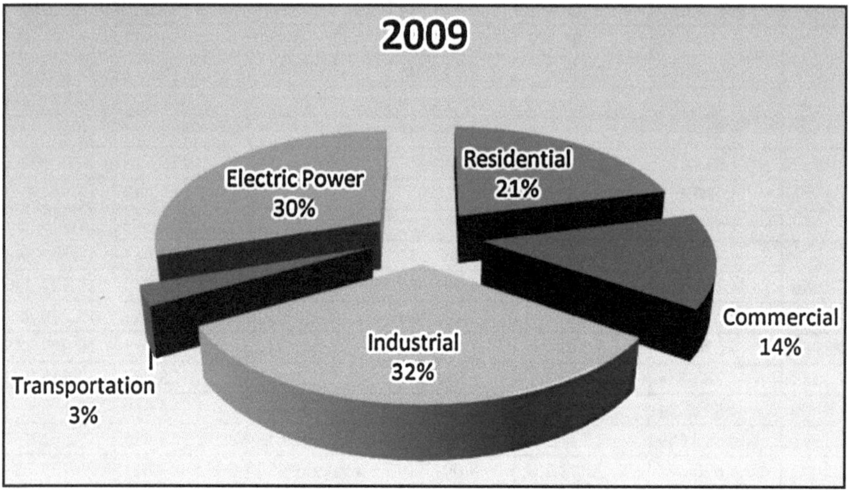

Reference: EIA/Monthly Energy Review, January 2011

Figure 7 compares sector-based natural gas consumption in 1950 and 2009, showing that industry accounted for 59% of natural gas consumption in 1950 but declined to 32% in 2009. Remarkably, natural gas is currently preferred by the electricity sector, in particular.[52]

Power plants are expected to account for a major share of the rise in natural gas consumption, with the electricity sector making up 59% of the increase. Moreover, GTL plants account for a small but increasing share of natural gas demand and are evolving into a new big market for natural gas by finding reserves that

51 IEA (2011); Energy Efficiency Governance, Handbook Second Edition, International Energy Agency Press

52 IEA (2011); Energy Efficiency Governance, Handbook Second Edition, International Energy Agency Press

are not distanced from current markets. Moreover, the use of natural gas is becoming commonplace in the residential and service sectors, resulting in a rise in natural gas demand from these consumer groups.[53]

Fig. 7: Share of Sectors in the World Natural Gas Consumption (1950-2009;%)

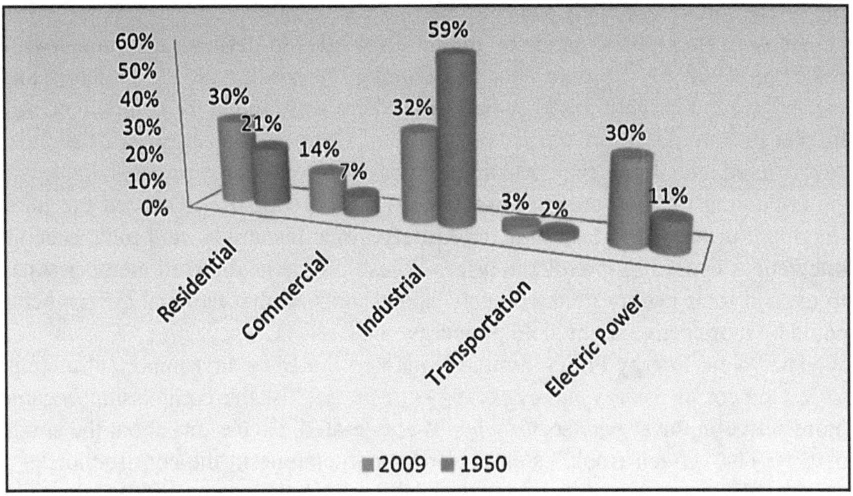

Reference: EIA/Monthly Energy Review, January 2011

Since it is one of the world's biggest economies, the European Union imports energy to meet its needs. The EU has adopted policies to support alternative projects in order to decrease its energy dependency. In this context, it aims to build an energy market that meets the criteria in terms of efficient regulation, long-term planning, supply security, environmental protection, and crisis management (Austvik, 2009, 90).

The EU's energy policies are as old as the EU itself and rely on three different treaties. The 1951 Paris Treaty and 1957 Rome Treaty show how much the EU cares about the energy sector. In this context, the EU's energy policy began with the coal policy of The European Coal and Steel Community (ECSC) and continued with the European Economic Community's (EEC) common policies for the nuclear sector. Both of these treaties aimed to compose free and integrated markets. They also contained clauses relating to petroleum, natural gas and electricity for building a common market.

53 IEA (2009); Energy Efficiency Governance, Handbook Second Edition, International Energy Agency Press.

No decisions were made about energy policies when the EEC was constituted. The need to collaborate on energy policies increased when petroleum prices started to soar. The European Union took its first steps towards energy policies back in 1964 when the member states accepted the report on energy related problems. The aims of the EU were stated in this report, and encompassed fair competition, supply safety, low-price supply and consumer's freedom of choice.

Another attempt was made by multiple societies to define a common energy policy in 1968 by multiple societies coming. According to the Commission, barriers were obstructing energy trade, and there was a need for common energy market that would ensure the secure supply of lowest cost energy. The Commission offered a common energy market providing lower energy prices,

Following the two energy crises in 1973 and 1979, the EU faced the three important problems of lack of supply safety, price instability, and trade account imbalances caused by petroleum prices. These problems directed member states to expand their energy policy objectives and tools so that national governments could be more active in the field of energy.

The "New Energy Policy Strategy" marked the EU's first important attempt to implement an energy policy in 1974. In the 1980s, the commission became more active in the energy sector when it accelerated discussions about the single market. The "Green Book" is an important cornerstone in the construction of a common EU energy policy, which defined the EU's long-term safe energy supply strategies based on following items:

- Uninterrupted supply of energy products,
- Affordable price levels for all customers,
- To respect environmental concerns and
- To take care of sustainable development

10. Natural Gas Market in the EU

The European Union has 2% of the world's total natural gas reserves, and 25% of the energy it consumes is natural gas. The main problem it faces is that while the production of natural gas is decreasing, the amount of consumption is increasing.[54] Tables 7 and 8 show the share of natural gas relative to total energy consumption by Eurogas members and the EU27. The table indicates that 27 EU members consumed a total 1.804 PJ of primary energy in 2008. Natural gas ac-

54 Kavalov B., H. Petric and A. Georgakaki (2009); "Liquefied Natural Gas for Europe-Some Important Issues for Consideration", *JRC References Reports*, European Commission (JRC). p. 5.

counted for 25% of total consumption, which shows that natural gas is the most important energy resource after petroleum.

Table 7: Primary Energy Consumption (PEC) in EUROGAS Countries and EU27 (2008 years)

Country	Natural Gas*	Total Primary Energy*
Belgium	15.5	58.6
Czech Republic	7.7	43.5
Denmark	4.1	20.1
Germany	73.1	341
Estonia	0.7	5.5
Ireland	3.5	16.3
Greece	3.4	32.5
Spain	10.5	142.3
France	39.6	274
Italy	70	192
Latvia	1.3	4.7
Lithuania	2.6	9.3
Luxembourg	1.2	4.7
Hungary	10.5	26.9
Netherlands	34.5	81
Austria	7.5	34.2
Poland	12.5	99
Portugal	4.5	24.7
Slovenia	0.9	7.6
Slovakia	5.1	18.2
Finland	3.5	33.4
Sweden	0.8	52.7
United Kingdom	94.9	224.3
Bulgaria	3.3	20.1
Romania	13.9	37.8
Switzerland	2.8	28
Turkey	31.7	97.3
EU 27	451.7	1804.7

** PJ-GCV*

Reference: EuroGas Annual Report, 2008-2009, pp.26

Fig. 8: Share of Primary Energy Consumption EU27 (2008)

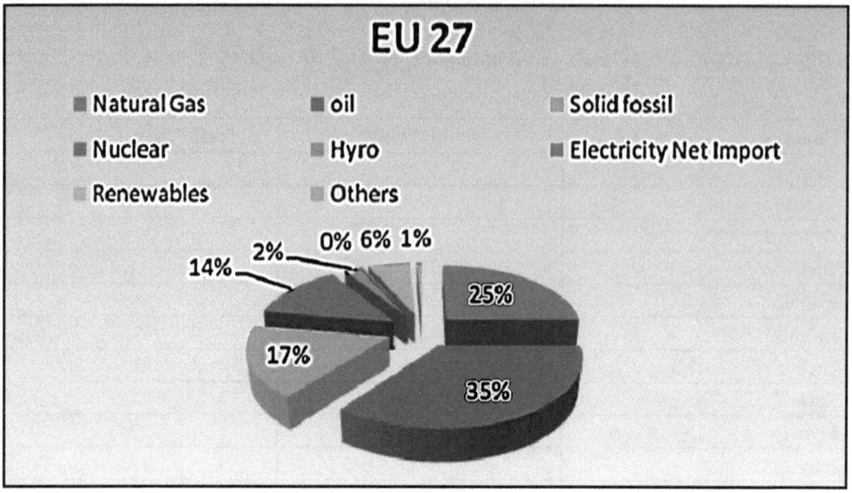

Reference: EuroGas Annual Report, 2008-2009, pp.26

Figure 9 shows the consumption of natural gas in the EU27 and proves that natural gas is the second most important energy source in the European Union. Of the total natural gas consumption, 31% is used in power plants, 26% in housing, 21% in the industrial sector, 9% in the service sector and the rest is consumed by other end users.[55]

Figure 10 shows the production and consumption of natural gas over the period between 2000 and 2008 in the EU27, and indicates a decrease in natural gas production and an increase in natural gas consumption.

55 European Commission (2009); Market Observatory for Energy Report 2009, European Commission Directorate-General for Energy.

Fig. 9: Consumption by Sector (EU27, 2007)

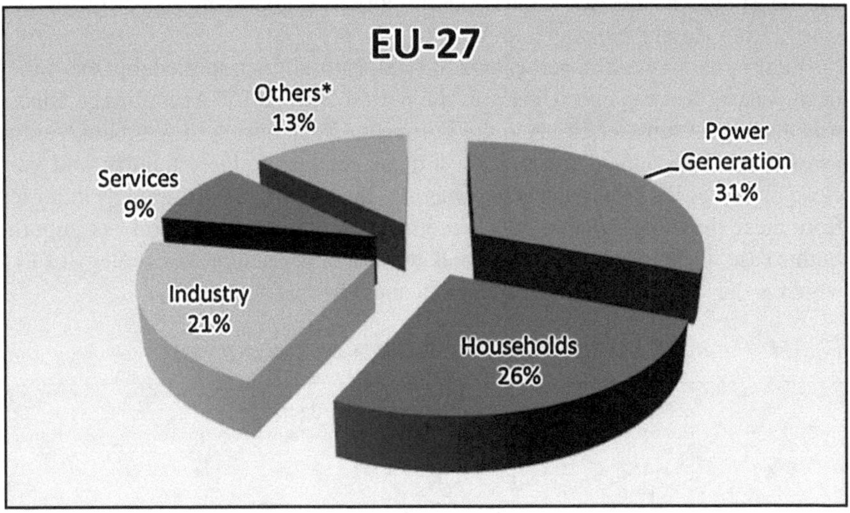

**others: 4% energy sector consumption, 3% final non-energy consumption (chemical industry), 2% final energy consumption other sectors, 1% agriculture, 3% other*

References: European Commission Report, 2009, pp.12.

Fig. 10: Natural Gas Production and Consumption (2008; 1000 tone)

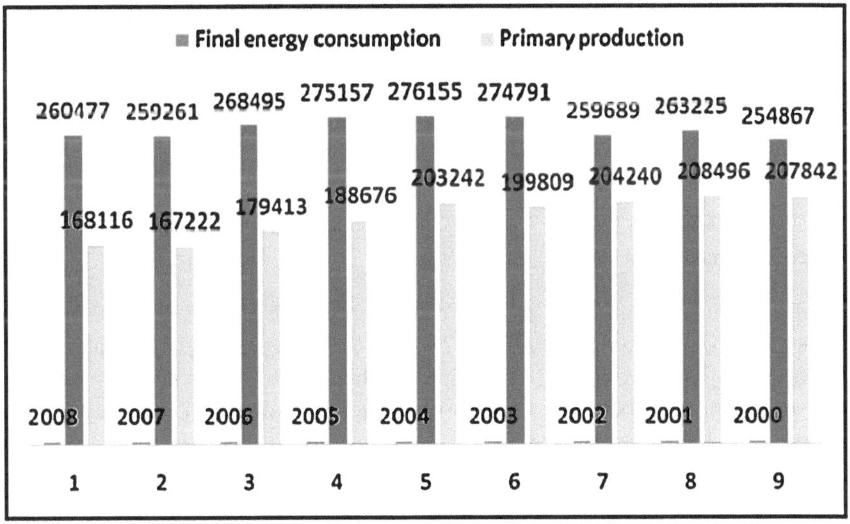

Reference: Eurostat.

Because of the difference between the amounts of natural gas that are produced and consumed in the EU, natural gas imports are important and originate primarily from Russia, Norway and Algeria.

Figure 11 shows the percentage of total natural gas imported by the EU27 from Russia, Norway and Algeria in the period 2002-2007. Accordingly, Russia was the largest supplier of gas to the European Union during the period. Over the years, Russia accounted for over 40% of all total natural gas imports, and very close to half of the total natural gas imports. In addition, the natural gas imported from these three countries in 2007 accounted for 85% of all natural gas imports. In this case, it can be claimed that the basic aim of the natural gas strategy of EU, which is the "diversification of supply," cannot be managed easily.

Fig. 11: Natural Gas Imports into the EU27 (% of total imports)

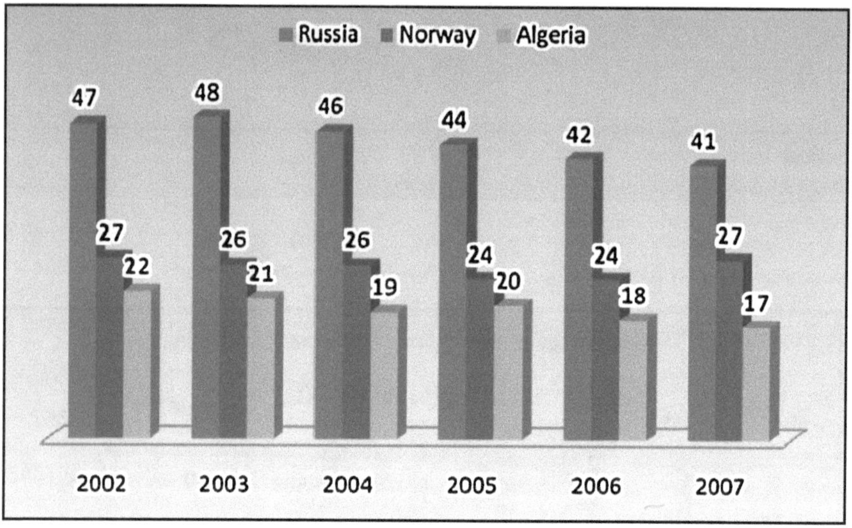

Reference: European Commission Report, 2009, p.42

Studies estimate that the energy consumption of the existing 27 member states will continue to rise until 2030 and that energy consumption in 2030 is expected to be approximately 15% higher than 2000. This increase in energy consumption is expected to meet with the increasing market share of natural gas and renewable energy sources. In other words, natural gas in Europe is estimated to be the fastest-growing fossil fuel over the next 20-year period. On the other hand, the number of power plants that use natural gas is predicted to more than double in 2030 compared to 2000. It is projected that the consumption of natural gas going to increase, partly because it is a clean, efficient and low-carbon fuel.[56]

Conclusion

The clean, cheap and safe supply of energy is absolutely crucial for the development of society. Countries will continue their R&D studies into cheap and clean kinds of energy production. Commercial, political and even military efforts are being undertaken in order to secure uninterrupted energy supplies. Besides, countries with poor energy resources are looking for solutions for efficiently using current and renewable energy resources, and are investigating which energy resources should be appropriately prioritized.

At present, numerous countries are trying to improve their strategies for sustainable development through sustainable energy. By making use of renewable energy resources, countries hope to lessen their energy dependency and revive their economies. Because of various difficulties surrounding the production and utilization of alternative energy resources, however, their share in total energy consumption is very low, and traditional energy sources and fossil fuels are still being used as the primary energy resources for now. Fossil fuels, consisting of petroleum, natural gas, and coal, have an important place among other energy resources. Research into alternative energy sources, in particular, and environmental sensitivity are making natural gas one of the most important energy resources. Natural gas ranks second in the EU and in the world as a whole. The EU's main aim in respect of natural gas is to obtain supply security. Resource variation is very important for supply security but the current situation does not lend itself to claiming that the EU has a rich energy supply market. The European Union imports 48% of its natural gas from Russia. In the natural gas market, which demonstrates monopolistic characteristics, some changes have started to emerge. The European Union first aimed for an integrated market, before liberalization and a competitive approach were adopted.

56 Hanore A. (2006); "Future Natural Gas Demand in Europe: The Importance of the Power Sector", *Natural Gas Research Programme*, No. 1/103, Oxford Institute for Energy Studies. p. 5.

References

Alam, M. Shahid (2006); "Economic Growth with Energy", MPRA *Paper*, No. 1260, Munich Personal RePEc Archive.

Alessandro, Ronjaglia (2003); "Energy and Market Power: Am Alternative Approach to the Economics of Oil", *Journal of Post Keynesian Economics*, 25 (4), pp. 641-659.

Ausvik, Ole G. (2009); "EU Natural Gas Market Liberalization and Long-Term Security of Supply and Demand", *Political Economy of Energy in Europe* (Ed.) G. Femann, pp. 85-115.

Balmaceda, M. M. (2002); "EU Energy Policy and Future European Energy Markets: Consequences for the Central and East European States", Forschungsschwerpunkt Konflikt- und Kooperationsstrukturen in Osteuropa an der Universität Mannheim.

Belkin, Paul (2008); "The European Union's Energy Security Challenges", *CRS Report for Congress*, Congressional Research Service.

BP (2010); BP Statistical Review of World Energy 2010.

Clawson P. and S. Henderson, (2005); "Reducing Vulnerability to Middle East Energy Shocks", *Policy Focus*, No. 49, The Washington Institute for Near East Policy.

EIA (2011); Monthly Energy Review: January 2011, U.S. Energy Information Administration (EIA) Press.

EUROGAS (2010); Natural Gas: The Natural Choice for Efficient Energy, the European Union of the Natural Gas Industry (EUROGAS) Press

EUROGAS (2009); Annual Report 2008-2009, The European Union of the Natural Gas Industry (EUROGAS) Press.

EUROGAS (2004); Natural Gas: The Energy for a Sustainable Future, The European Union of the Natural Gas Industry (EUROGAS) Press.

European Commission (2009); Market Observatory for Energy Report 2009, European Commission Directorate-General for Energy.

European Commission (2007); Communication from the European Council and the European Parliament: An Energy Policy for Europe, *Commission of the European Communities*, Sec (2007) 12, Brussels.

European Commission (2000); Opening Up to Choice: Launching the Single European Gas Market, European Commission, Directorate-General for Energy and Transport, Unit for Internal Market, Public Service, Competition and Application of Community Law Published

Goldember, J, et al., (2001); "Energy for the New Millennium", *Ambio*, 30 (6), pp. 330-337.

Hall, C., et al. (2001); "The Need to Reintegrate The Natural Sciences with Economics", *BioScience,* 51(8), pp.663-673.

Hanore A. (2006); "Future Natural Gas Demand in Europe: The Importance of the Power Sector", *Natural Gas Research Programme*, No. 1/103, Oxford Institute for Energy Studies.

IEA (2010); Energy Efficiency Governance, Handbook Second Edition, International Energy Agency Press.

IEA (2009); World Energy Outlook: Executive Summary, International Energy Agency Press.

IEA (2008); The European Union 2008, IEA Energy Policies Review, International Energy Agency Press.

Imran, K and M. M. Siddiqui (2010); "Energy Consumption and Economic Growth: A Case Study of Three SAARC Countries", *European Journal of Social Sciences*, 16(2), pp. 206-213.

Kavalov B., H. Petric and A. Georgakaki (2009); "Liquefied Natural Gas for Europe-Some Important Issues for Consideration", *JRC References Reports*, European Commission (JRC).

MIT (2010); "The Future of Natural Gas", *Interim Report*, Massachusetts Institute of Technology (MIT) Press.

Nordhaus, William D. (1973); "The Allocation of Energy Resources*", Brookings Papers on Economic Activity, No.* 3, pp. 529-579.

Paul S. And R. H. Bhattacharya (2004); "Causality between Energy Consumption and economic Growth in India: A Note on Conflicting Results", *Energy Economics*, 26(2004), pp. 977-983.

Rees R. and S. Scholz (2010); "Electricity Market Design for Germany", Sosyo-Ekonomi, 6(12), Enerji Özel Sayisi, pp. 39-62.

Stern, David I. (1999); "Is Energy Cost an Accurate Indicator of Natural Resource Quality?" *Ecological Economics*, 31, pp. 381-394.

Stern, David I. (1993); "Energy Use and Economic Growth in the USA, Multivariate Approach", *Energy Economics*, 15, pp. 137-150.

Stern, D. I and Cleveland, C.J (2004) "Energy and Economic Growth", *Rensselaer Working Papers in Economics* 0410, Rensselaer Polytechnic Institute, Department of Economics.

Özsabuncuoğlu İ. H. And A. Uğur (2005); Doğal Kaynaklar: ekonomi, Yönetim ve Politika, İmaj Yayinlari, Ankara.

Toman, M. and B. Jemelkova (2003); "Energy and Development: An Assessment of the State of Knowledge", Resources for The Future Discussion Paper, 03-13.

Truddea, N. and I. Murray, (2011); "Development of Energy Efficient Indicators in Russia", *IEA Working Paper*, International Energy Agency Press.

TÜBİTAK (1998); Enerji Teknolojileri Politikasi Çalişma Grubu Raporu-Yönetici Özeti-Alt Grup Önerileri Toplami, TÜBİTAK Vakfi Yayinlari, Ankara.
Yorkan Arzu (2009); "Energy Policy of the European Union and Its Effects on Turkey", *Bilge Stratejisi*, 1(1), pp. 24-39.

Increasing Energy Demand of the EU and Turkey's Role in Supplying Energy to the EU: Analyzing the Mutual Costs and Benefits[1]

Abdulkadir Develi, Ahmet Ay and Zeynep Karaçor

Energy demand is increasing with quick modernization life standard of world and increasing energy demand of EU, and United States is also another issue to be considered. Especially, economic development of China, Latin America and India are important factors in terms of global energy demand. EU does not have adequate energy resources when it is compared with its grant amount of energy consumption. North Sea, being, under the control of England, has 4.4% of total world energy resources, but cost of oil production is three times more than Middle East energy costs. High energy costs, inadequate energy resources and growing energy demand makes EU as one of the most important energy consuming region. Russia is one of the most important traditional suppliers for the EU and is trying to maintain its position in the future. But energy cooperation between Russia and EU contains some problems of competition and energy security. Lack of adequate infrastructure is a big problem for Russia to meet growing demand of EU, and Russia sometimes use energy as political tools as it is seen in the example of Ukraine Crises. So, diversification of energy resources of EU will provide more competitive energy supply and strong energy security of EU. Therefore, EU needs to regional cooperation for energy diversification. Caspian region is an energy rich region. Therefore Turkey's strategic location makes it a natural "Energy Bridge" between the major oil producing areas in the Middle East and Caspian Sea regions on the one hand, and consumer markets in Europe on the other. So, energy cooperation with Turkey with EU is hot topic in EU energy policy.

This paper tries to examine cost and benefit analyses between Turkey and EU in the aspects of energy cooperation. These cost and benefit analyses also include projects that are under the construction or still project process. EU has been providing diversification of energy resources for more competitive energy supply and reducing energy dependence to Russia. Turkey has some benefits from this process. Because, pipelines that provide energy to Europe passes though Turkish boundaries, and Turkey collect transportation fee from these pipelines and also EU supports these project financially. This also contains cost for EU.

1 This chapter is the extended version of the paper that was presented at the International Conference on Eurasian Economies, Istanbul, November 4-5, 2010.

1. Introduction

Energy has played a crucial role for nations since the beginning of history. Countries compete against each other to reach the energy resources first. Energy is a primary agent in the development of wealth and a significant factor in economic development (Develi & Karaçor, 2007; 2). Global energy demand is increasing as the modernization of the standard of living in the world quickens. The growing energy demand of the EU and United States is another point to be considered. Economic development in China, Latin America and India is a particularly important factor that influences global energy demand. A social state is measured in terms of both individual income and individual energy consumption. Energy is an economic development tool for nations.

Therefore, energy is an important political issue, both for nations and for the EU. Since the last Ukrainian crisis, the EU has been trying to compose a more extensive common energy policy throughout the territory. Back then in 2006, the EU witnessed a "9/11" event when the security of European gas supply became a very topical subject following the cuts in Russian supplies to the Ukraine in the first days of the year, which had the consequence of restricting the availability of supplies to some European countries (Stern, 2006; 3). Since then, the EU has been trying to diversify its energy resources. Diversification of energy resources involves the same cost and benefit aspects as the economy and politics. Because if the EU is to obtain alternative energy resources it needs political and economic stability, and political and economic stability can only be assured by cooperating with transit and producing countries. So, the EU is trying to use all its authority in current and future bilateral negotiations and agreements, and offering balanced, market-based solutions, first of all to its traditional suppliers, but also to other major producing and transit countries.

Moreover, the EU has been researching different alternatives, such as important producers in the Black Sea, Caspian, Middle East and Gulf regions. Turkey is therefore becoming a crucial energy hub for supplying energy requirements from producer regions, and this is strategically important for EU energy security. On the other hand, cooperations on pipeline projects, such as the Nabucco Project and the South European Gas Ring Project (SERG), are very big opportunities for both the EU and Turkey in terms of strengthening relationships with the EU for future membership (Green Paper, 2006;1). In terms of cooperation between the EU and Turkey, the establishment and development of trans-European (energy) networks and the promotion of proper interconnections and the interoperability of national networks aim to take full advantage of the internal market (European Parliament, 2006; 3). The strategic potential offered by Turkey makes it a huge energy hub for the EU. Transporting gas and petrol from Russia, the Caspian Sea region and the Persian Gulf through Turkey by ships and tankers involves mutual costs and benefits for the EU and Turkey. These costs and benefits can be calculated in political terms, as an energy security concept and economic cost and benefit, such as the aspects of pipeline construction and transportation fee for the EU and Turkey. In this paper, natural gas and oil are taken into consideration among other energy resources but natural gas is more important than oil because demand in Europe is higher for natural gas than for oil. Transporting natural gas is more restricted than oil. Oil can be transported by pipeline and tankers but natural gas can mostly only be piped, because the transportation of natural gas as LNG (Liquid Natural Gas) is too expensive. As such, natural gas pipelines are becoming increasingly important for the EU.

2. Analysis of the General Energy Demand of the EU

Compared with its enormous energy consumption, the EU does not have adequate energy resources. Its dependence on energy imports will increase steadily from 2010 to 2020. The EU's dependency is growing daily. As a whole, it does not have the necessary resources to cover its domestic demand for energy (Bacik, 2006; 294).

The European Union has coal, oil, and natural gas reserves. There are six oil producers in the European Union, primarily in North Sea oilfields. The United Kingdom is the largest producer by far; however Denmark, Germany, Italy, Romania and the Netherlands also produce oil. According to 1997 data, the North Sea region that is under the control of England has 4.4% of the total world energy resources, but oil production costs three times more than in the Middle East. High energy costs, inadequate energy resources and growing energy demand make the EU one of the most important energy regions. European gas demand will rise from presently 540 billion cubic metres (bcm) to around 800 bcm in 2030, rendering Europe the largest gas market in the world. Oil demand will increase, too, from the present 16 million barrels per day (mbd) to 20 mbd in 2030. At the same time, generic European resources in the North Sea have peaked and its contribution to future European energy supplies will be sharply reduced (Goldthau, 2008; 2).

The source dependence of OECD Europe will increase from 36% in 2002 to 69% in 2030, and raises the legitimate concern of technical and political reliability. The bulk of the supply will be concentrated on Russia 33%, North Africa 27% and the Middle East 17% (Weisser, 2007; 2).

The structural peculiarities of natural gas and the dependency on specific regions – and especially Russia – make it more important in terms of security of supply. The Ukrainian crises severely affected European gas supply. The Ukrainian pipeline system is pivotal in supplying gas to Europe and the consequences for Europe's gas supply security were quick to follow: from January 1-3, Gazprom's gas supply to France decreased by 25-30%; supply to Austria decreased by 33%; and Italy received approximately 25% less gas than normal (Spanjer, 2007; 1). However, the EU consumes twice as much natural gas as petrol; it is therefore more important to ensure a secure gas supply to the EU and to implement alternative projects to diversify gas supplies than to assure oil security for the EU. The EU accounts for 17% of total world natural gas consumption. It is estimated that the petrol dependence of the EU will reach 70% by 2030 and natural gas dependence will be 80% about 25 years later. The EU has so far imported 90% of its natural gas demand from specific regions and countries (EC, 2000;19). The EU's energy imports come from specific regions; for example, 40% of the gas is imported from Russia and 45% of its petrol is imported from OPEC countries (Morelli, 2006; 9). Natural gas was the fastest growing fuel among the fossil fuels used in the EU, increasing by 2.78% per year between 1990 and 2005 (EU, 2008; 72).

The Middle East and North Africa (MENA) are among the leading countries that supply energy to the EU. Although Russian reserves are larger than any single MENA country, many of the latter countries have reserve to production ratios that exceed 100 years, suggesting ample potential for exports. MENA countries have always been seen as a huge potential import resource for European gas markets (Stern, 2006: 3).

Table 1: Middle East and North African Gas Export Projections 2003-2030 (bcm)*

	TO EUROPE			TOTAL EXPORT			
	2003	**2010**	**2030**	**2003**	**2010**	**2010**	**2030**
Middle East	2	35	117	34	102	185	244
North Africa	61	83	170	63	86	143	200
TOTAL	63	118	287	97	188	327	444
Major Exporters**:				**2003**	**2010**	**2020**	**2030**
Qatar				19	7	126	152
Algeria				64	76	114	144
Iran				–	5	31	57
Egypt				–	10	19	28
Libya				1	2	13	34
Iraq				–	1	7	17
TOTAL				84	172	310	432

* *in addition to the countries listed, MENA includes: UAE, Kuwait and Saudi Arabia.*
** *figures are for 'net trade'.*
Source: International Energy Agency, World Energy Outlook 2005, Paris: OECD, 2005, pp. 178-604.

On the other hand, the EU has to pay huge amount of money for energy. The PRIMES model includes a detailed calculation of energy related costs and attributes these costs to the end consuming sectors. Cost analysis includes expenses for purchasing energy products, annuity payments corresponding to investment in end-use equipment, operating expenses and spending to improve energy efficiency (e.g. insulation, etc.). The end-user prices of energy products are also estimated by the model on the basis of a detailed cost analysis of supply, which includes import prices of energy products, extraction costs, distribution costs and annuity payments for investment. The determination of end-use costs takes the view that prices reflect total costs plus profits. The total energy related costs increase by 1.85% per year, between 2005 and 2030. This growth is lower than GDP growth, implying that energy costs as a percentage of GDP decrease in the baseline scenario.

Table 2: Energy Costs of the EU

	2000	**2005**	**2010**	**2020**	**2030**
Total Cost Related to Energy in billion €	978	1080	1219	1516	1709
As% of GDP	9.73	9.87	9.81	9.66	9.15

Source: European Commission, 2008 'Directorate-General for Energy and Transport' Luxembourg: Office for Official Publications of the European Communities.

As Table 2 shows, energy costs in the EU account for 9-10% of its GDP. The EU imports most of its energy demand, and its energy costs will rise as its natural gas consumption increases. This increase is mainly caused by the price of natural gas, which rises more than the average energy price. The prices of oil products follow the progress of world oil prices (EU, 2008; 77). Therefore, energy diversification by the EU will provide different energy from different producer's countries, which will create a more competitive market structure.

3. Role of Turkey in the Energy Demand of the EU

Turkey has historically been of great geopolitical importance, both to states in the surrounding regions and to more remote countries. Turkey has regained its geopolitical importance for other reasons. One of the most important reasons is its location between large energy markets and major energy producers (Tekin &Walterowa, 2007; 1). Turkey's strategic location makes it a natural 'Energy Bridge' between the major energy producing areas in the Middle East and Caspian Sea regions (Devlet, 2005; 1). Turkey is strategic region, which is crossed by

71.8% of the world's energy sources (BP, 2005). As indicated by John Roberts, the transmission of natural gas is a regional issue and it reinforces cooperation between countries in this region. The EU is trying to diminish its energy dependency on a specific country and region by diversifying its sources from the Caspian Black Sea and Mediterranean. Turkey is an important link in the East-West energy corridor, channelling Caspian and Middle Eastern energy to European and world markets. The Baku-Tbilisi-Ceyhan and Iraq-Ceyhan oil pipelines, Turkmenistan-Turkey-Europe, the Shah Deniz, the Black Sea and Turkey-Greece natural gas pipelines would provide the EU with alternative routes for energy supply that could be used to provide energy security to the EU.

Turkey is situated at the junction of three continents (Asia, Europe and Africa) and stands as a bridge between Asia and Europe (Demirtaş, 2000;15). The country is located in south-eastern Europe and south-western Asia. Turkey's strategic location makes it a natural energy bridge' between the major oil producing areas in the Middle East and Caspian Sea regions, on the one hand, and the consumer markets in Europe on the other. According to a report issued in 2006 by the EU on the contribution of Turkey to the EU in terms of energy, Turkey will facilitate the EU's access to central Asia, the Caspian region and the Middle East and be a transit country for transporting energy sources to the EU.

In terms of providing energy security for the EU, it is worth distinguishing between oil and natural gas, because it is more difficult to secure the supply of natural gas compared with oil, due to the route inelasticity caused by the necessity to transport natural gas by pipeline and to the high cost of LNG terminals.

The secure supply of oil is a global issue but, since the supply security of natural gas is a more regional problem, relations between neighbours and cooperation are a basic element (Roberts, 2004; 28) Dependency on oil imports can be met by refineries, different ports and tankers, but if any country or region depends on one country or region, it will be more difficult to diversify gas supply sources because of its structural restrictions. In this respect, Turkey also acts as a bridge for producing countries to Europe.

Azerbaijan, Kazakhstan and Turkmenistan are landlocked and have to cooperate with their neighbours to export energy. Pipeline construction requires foreign investment, infrastructure and employment. However, the transmission fee for the transit country is also important. All of these reasons make it clear how important sensitive negotiations are in pipeline construction.

There are currently four cross-border gas pipelines through Turkey (Özdemir, 2008):

- From the west, the Russia-Turkey western pipeline, carrying Russian gas, with a capacity of 14 billion cubic metres (bcm) per year;

- From the north, the Blue Stream pipeline, carrying Russian gas, with a capacity of 16 bcm per year;
- From the east, the South Caucasus pipeline, carrying Azerbaijani gas, with a capacity of 8 bcm per year presently, but upgradable to over 20 bcm;
- From the east, the Tebriz-Erzurum pipeline, carrying Iranian gas, with a capacity of 20 bcm.

Table 3: Supply Potential to the EU via Turkey

Country	Volume	Transit Country	Potential by 2015	Existing System
Iran	10 bcm	Turkey	20-30 bcm	3-10 bcm
Turkmenistan	13 bcm	Iran/Turkey	30 bcm	13 bcm
Turkmenistan	16 bcm	Aze.Geo/Turkey	30 bcm	None
Saudi Arabia	10-20 bcm	Jordan/Syria/turkey	20 bcm	None
Azerbaijan	8 bcm	Turkey	20 bcm	8 bcm
Iraq	10 bcm	Turkey	10 bcm	8 bcm
Egypt	4 bcm	Jordan/Syria	10 bcm	Link to Syria

Sources: Roberts (2004)

According to Roberts (2004), Turkey's role as a gateway through which gas can enter the European Community is becoming increasingly important as the European Union grapples with the interrelated geopolitical position. Many natural gas pipelines have been established and some of them are still in the project development phase (Roberts, 2004: 11). But two of these projects – the BTC and Nabucco pipelines – are more important than others because of the huge amount of energy they could potentially supply to Europe.

Baku-Tbilisi-Ceyhan (BTC) Crude Oil Pipeline: This project aims to transport crude oil produced in the Caspian Basin, in places such as Azerbaijan and Kazakhstan, by pipeline to a marine terminal on the Mediterranean coast of Turkey in Ceyhan and then to the world markets by tankers (Babali, 2005: 30). Within this framework, a major aspect of Turkey's energy strategy is to complete "the East-West Energy Corridor". Total capacity of the 1768 km long pipeline is 50 million tons of oil per year (or 1 million barrels a day) (İpek, 2006: 2). The pipeline will have an approximate length of 1,100 miles (1,770 km) and will cost an estimated $ 3.6 billion. Seventy percent of the project (about $2.6 billion) will be funded by credit. It was initially planned for completion in 2005; however, due to some delays, the first oil was expected to flow through it in the second half of 2006 (Babali, 2005: 46). Turkey's role in this project is very critical in respect of providing pipeline security. This project will generate revenue of 600 million dollars per year for Turkey.

Nabucco pipeline project aims to connect the large gas reserves in the Caspian Sea and the Middle East across Turkey's borders to European markets, including the territory of five Nabucco partners, namely Turkey, Bulgaria, Romania, Hungary and Austria. These Nabucco partners have agreed on the commercialization of the Nabucco pipeline capacity and on maintaining and operating the infrastructure, which will be the responsibility of national companies. A Nabucco Gas Pipeline International GmbH has formally applied for exemption based on Article 22 of Directive 2003/55/EC of 26 June 2003 concerning the common rules governing natural gas for the internal market (Unece Gas Centre Report, 2005).

The Nabucco project was first discussed in February 2002 between OMV gas and BOTAS. Eventually, in October 2002, a cooperation agreement was signed between BOTAS, Bulgar gas, MOL (Hungarian Oil Gas plc.), Trans Gas and OMV Gas, whose main goal was to conduct a feasibility study for the construction of new gas pipeline (BOTAŞ, 2009). In December 2003, a major agreement was signed between OMV Gas and the other four partners as associated beneficiaries, and the European Commission. Finally, five companies, namely BOTAS (Turkey), Bulgargaz (Bulgaria), Transgaz (Romania), MOL (Hungary) and OMV Gas (Austria) launched a study for a pipeline to transport natural gas from points on the eastern border of Turkey, through Bulgaria, Romania and Hungary to facilities at Baumgarten in Austria.

The total eligible costs of the study will be supported by the EC. It is estimated that the project will cost 4.6 billion Euros (The Business, 2007). Obviously, the EU's motivation in financing a crucial amount is to achieve its own security of supply, and the position of Turkey is very important in terms of the realization of this project. Compared with Russia, it is estimated that 30 billion cubic metres of gas will be supplied to Central Europe by 2015. All technical, legal, commercial and financial issues will be covered during the development phase. As far as the technical issues are concerned, the basic and detailed engineering analysis will be performed, in principle, to meet all the requirements for environmental impact assessment in the Nabucco countries and to obtain all approvals from the respective authorities.

The South European Gas Ring Project (SERG) 'Turkey – Greece – Italy interconnector project' will provide another alternative source of energy to the EU. This project aims to pipe gas to Europe from the Caspian and Middle East gas reserves. Turkey plays a key role in this project, too (EC, 2003).

The Turkish portion of the project cost is 135 million Euros, which will be 50% financed by the EU and 50% by BOTAS. The EU is providing financial and economic support for the project. Turkey will be entitled to taxation revenue from this project based on border length, and will therefore earn the largest transportation fee. In this respect, Turkey will be entitled to 60% of the taxation revenue from the pipeline, which will earn Turkey estimated income of approximately 400-450 billion Euros (Çerçi, 2009:17).

The implementation of the 'Turkey – Greece – Italy interconnector' will offer many advantages for all the involved parties. For Greece, it will be an important contribution to the diversification of gas supplies, and will accelerate the liberalization of the gas market according to free market economy principles. It will also enhance the cooperation with Turkey and the producing countries. Turkey, together with Greece, will play an important role as a transit country of Eastern gas to Europe. When we look at the aspects of the producing countries, this project offers significant opportunities for their gas flow to Europe. Finally, Italy and the other European countries will be interconnected, thus securing supply and increasing the diversification of supply in a period where Russia's energy policy is proving ambiguous. BOTAS, DEPA and EDISON GAS are willing to realize this ambitious project and many important decisions will be made in this direction over the coming months.

In 2030, the Nabucco project will provide more energy supply to Europe than any other energy route (North Africa, North Sea and its domestic production) according to statistics issued by BP in 2006. The Nabucco Project will probably be the world's largest natural gas pipeline, offering huge potential in 2030.

The economic and political costs and benefits of the energy cooperation between the EU and Turkey are as follows:

- Easy and safe access will be assured to the consumption centre in Europe with the existing transmission infrastructure and a cost-based transportation tariff on transit pipelines.
- Investment by European gas companies in exploring, generating and creating potential new business opportunities in connection with these projects.
- Ensuring the long-term energy supply to the EU will relieve the energy markets and generate visible price advantages for the EU, which pays a fortune for its energy.
- Increasing bilateral trade capacity, establishment of economic interdependence and strong political relations with the Middle East and Caspian regions will be achievements of these projects.
- Contribution to supply security by creating diversification of supply routes; dependence of consuming countries on particular region will be reduced by these projects.
- European companies will have opportunity to gain a foothold in a very fast-growing market like Turkey.

- Turkey has joined the pipeline projects, funded pipeline construction costs and will collect transportation fees from these projects.
- Turkey ensures the security of these pipelines.
- The EU energy partnership with Turkey is strengthening its position in its attempt to join the EU.
- Oil tankers passing through the Straits pose an environmental risk. The new pipeline project will mitigate this risk.
- The EU has taken over a huge proportion of the cost of these projects.
- Turkey will benefit from these pipeline transportation projects by ensuring the security of supply more effectively and securing investment partnerships for a high capacity transportation infrastructure.
- The projects will create new infrastructure for inaccessible potential suppliers to transport additional energy.
- Turkey will benefit in future from decreasing prices in the long term, collecting transportation fees, making the most of the unused capacity of the existing transmission infrastructure, creating export opportunities to overcome possible take-or-pay problems.

Conclusion

The EU is trying to diversify its energy resources by cooperating with producing and transit countries. The EU's search for alternative energy routes to Russia involves both costs and benefits. Building new pipelines is hugely expensive, and the EU is covering most of the cost, such as the Nabucco project. Despite the high cost of new pipeline construction, the EU gains so many advantages from these projects, which can create a more competitive energy market structure by eliminating the monopoly of Russian energy company GASPROM in European markets. Energy security is another reason for the diversification attempts of the EU following the last Ukrainian crisis, which is still a hot topic on the EU's agenda. New pipeline projects definitely necessitate strong cooperation between bordering countries. Turkey sees itself as the energy bridge for countries transporting to Europe. The energy cooperation between the EU and Turkey also generates some costs and benefits for Turkey. Turkey is directly involved in some pipeline projects, and is partially funding them. Moreover, Turkey has to secure the pipelines. On the other hand, numerous petrol tankers cross the Bosporus, posing a huge environmental risk. Turkey will benefit from the transportation fees and gaining membership to Europe.

The EU's financial cooperation instruments should be mobilized in full to promote the restructuring and development of the energy sector in partner countries, regional cooperation, infrastructure interconnections, new pipelines, energy efficiency, and renewable energy sources for their mutual benefit. Member states need to support the ongoing bilateral and regional energy cooperation financed by the EU, the European Investment Bank, the European Bank for Reconstruction and Development and other international financial institutions. After all, the EU's energy costs make up 8-10% of its GDP, which virtually equates to the value of EU exports.

In light of the natural gas projects that Turkey is involved in, and the liberalization process of our energy market with special focus on energy security, Turkey is a crucial partner for Europe with regard to energy supply diversification. New projects will offer new advantages for FDI. In other words, our objective to become Europe's fourth main source of energy supply, after Norway, Russia and Algeria, overlaps the EU's energy security policy and opens up a new area for cooperation between Turkey and the EU.

Cooperation projects, such as NABUCCO and SERG, will rapidly align Turkey with EU energy standards. Turkey has the opportunity to develop regulations for the natural gas market to provide a more efficient market and improved infrastructure services for NABUCCO, SERG and BTC

References

Bacik, G. (2006), Turkey and Pipeline Politics, **Turkish Studies**, Vol. 7, No.2, 293-306, June 2006.

Babali, Tuncay, 2005. ''Implications of the Baku – Tbilisi – Ceyhan Main Oil Pipeline Project'', *Perceptions*, Winter.

British Petroleum, June 2005. *BP* **Statistical Review of World Energy**. London.

DEVELİ, Abdulkadir & KARAÇOR Zeynep, 2007, Economic Regulatory Reform on the Natural Gas Market in Turkey: Accession to the European Union and Future Gas Projects, **Insight Turkey**, *Vol. 9* Number 2.

Demirtaş, A. (2000) , Recent Advances in Biomass Conversion Technologies, **Science and Technology**, Sayi 6.

Devlet, N. (2005), Turkey's Energy Policy in the Next Decade, **Perceptions,** Winter 2004-2005.

Demirbaş, A. (2000), Recent Advances in Biomass Conversion Technologies, *Science and Technology*, Sayi 6.

European Commission, (2000), Green Paper: Towards a European Strategy for the Security of Energy Supply. COM 769 final. Brussels.
European Commission, (2008), **Directorate-General for Energy and Transport,** Luxembourg: Office for Official Publications of the European Communities.
İpek, P, (2006) , The Aftermath of Bakû–Tbilisi–Ceyhan Pipeline: Challenges Ahead for Turkey, **Perceptions,** Spring 2006.
Morelli, Vince L, (2006), The European Union's Energy Security Challenges, Analyst in European Affairs Foreign Affairs, **Defense and Trade Division**.
Özdemir, S, (2001), Avrupa Ekonomik Topluluğu, Avrupa Topluluğu, Avrupa Topluluklari, Avrupa Birliği, Roma Anlaşmasi, Avrupa Tek Senedi, Avrupa Birliği Antlaşmasi, Maastricht Anlaşmasi, Amsterdam
Anlaşmasi, Nice Anlaşmasi, www.dpt.gov.tr/DocObjects/Download/.../Kavram laraIilskinAN.pdf
Tekin, A & Walterova, Iva, (2007), Turkey's Geopolitical Role: The Energy Angle, *Middle East Policy,* Vol 14, No1 Spring.
The Business, (2007), Capital of Nabucco, September 20-28, 2007.
Stern, Jonathan, (2006), The New Security Environment For European Gas: Worsening Geopolitics and Increasing Global Competition for LNG, Oxford Institute for Energy Studies, October 2006.
Unece Gas Center, December, (2005), **Report on The Gas Company and Market Structure in The European Part of The Unece Region**, Geneva.
Roberts, John, (2004), The Turkish Gate: Energy Transit and Security Issues, **Centre For European Studies, EU–Turkey Working Papers**, No. 11.
Weiner, Myron, (1995), Nations: The Problem of Political Development, **Annuals of American Academy of Political and Social Science,** Vol. 358, March 1995.
UNECE Gas Center, December, (2005), **Report on The Gas Company and Market Structure in The European Part of The Unece Region**, Geneva.
BOTAŞ, (2009), http://.botaS.gov.tr/icerik/eng/projeler/yurtdisi.asp. 11.11.2009.
http://ec.europa.eu/dgs/energy_transport/international/regional/euromed/energy/doc/ conference_4/2003_11_20_annexes_en.pdf.

Energy Security of the EU: Turkey and Russia[1]

ŞabanÇaliş and AbdulkadirDeveli

1. Introduction

The importance of energy is accepted almost universally. The past demonstrates that a strong relationship has always existed between economic activity and energy. On the other hand, energy is key for an economic system and is widely used in the industry, residential and transport sectors. Therefore, low energy prices are the basic condition for the overall competitiveness of an economic system[2]. Natural gas is the most important energy source for the future. The abundance of natural gas, worldwide as well as domestically, coupled with its environmental soundness and multiple applications across all sectors, means that natural gas will continue to play an increasingly important role in meeting the world's demand for energy[3]. Demand for natural gas has been increasingly strong as it offers certain features that petrol cannot, and it has meanwhile become a rival to petrol. Demand has increased, not just for use directly as natural gas, but also for use in electricity generation. Interest in natural gas has increased around the world as it is specific to certain places. Developed countries, such as the European Union, USA and China, who all consume huge amounts of energy, are extremely interested in natural gas.

However, the need for energy security with regard to natural gas is just as intense as it is for petrol. Until 1973, there was no effective prediction by the world's petrol producers and politicians of a petrol crisis. Yet, the 1973 petrol crisis showed that limited petrol reserves couldn't produce limitless petrol as demanded. Henry Kissinger, recounting his experience at the time, said that the world had been caught unprepared by the petrol crisis in 1973 and he also emphasized that the same problem could arise with natural gas. According to Kissinger, the dimen-

1 This chapter is based upon the paper that was presented at the International Conference on Globalization, Energy and Environment, Warsaw School of Economics, 29-30 May, 2008.

2 Andrei, B. (2003)'New Dimension of Energy Security of The Enlarging EU and Their Impact on Relations With Russia' European Integration, December. Vol. 25 (4), pp. 351-369.

3 Kiliç, Ali, (2005) 'Turkey's natural gas necessity, consumption and future perspectives', Energy policy, Vol. 1, p. 3.

sion of gas security was more serious since natural gas and electricity are related closely to each other[4].

The intense fluctuation of oil and natural gas prices reflects the imbalance between global supply and demand. It also raises concern over energy security[5]. EU demand of natural gas is increasing and it depends almost entirely on third countries. This dependency on third countries, and especially Russia, has prompted the EU to try and create a more balanced supplier structure. The European Union is consistently striving to ensure that its external energy policy provides guarantees in terms of supply security. Over the past decade, relations between the EU and Russia have encompassed a broad set of policy issues. Among them, the energy chapter has played a crucial role. Indeed, Russian oil and gas exports are achieving ever-increasing importance as part of the country's foreign policy. The EU cooperates with Russia on energy. Russia has covered 25 percent of EU demand[6]. But energy cooperation with other third countries remains the EU's top priority because of security of supply. So, the EU has been searching for alternative important producing areas such as the Mediterranean, Black Sea, Caspian, Middle East and Gulf regions. Nowadays, the concept of security supply is gaining in importance for the EU. It can be described as ensuring easy access to energy sources and long term supply sustainability.

Turkey is becoming a crucial energy hub for supplies from the producer regions and its role is gaining in importance. This process also shows that the relationship between the EU and Turkey will be beneficial for the latter's future EU membership.

The future of the relations between Turkey and the European Union or the process of becoming a full member of the EU is going to be shaped and governed mainly by the geopolitical location of Turkey. These advantages mean that Turkey will play a crucial role for the EU in securing its alternative supply. The strategic advantages of Turkey are therefore giving it a chance to speed up accession to the EU by providing the European Union with alternative energy routes. However, cooperation on the Energy Community Treaty could also speed up this process. Cooperation on pipeline projects, such as the Nabucco project and South European Gas Ring Project (SEGR), give Turkey the opportunity to secure its relationship with the EU and make a significant contribution towards accelerating its accession to the EU.

---

4 IEA, 2004, *Security of Gas Supply in Open Markets,* http://www.iea.org/textbase/nppdf/free/2004/security.pdf.

5 Gawdat, B (2005) 'Energy Security: The Caspian Sea', Minerals and Energy, 2:3-15.

6 EU Commission Report: External energy relations: From principles to action (Com. Report. 560 final), Brussels, 12.10.2006, p. 4.

The project has been put forward for transferring combined Caucasian and Middle East natural gas to Europe through Turkey. The project, which aims to transfer nearly 30 billion cubic metres of natural gas to Europe annually, was launched in 2002, and moved forward by a consortium constituted under the leadership of Austria[7]. The NABUCCO pipeline, which was expected to supply gas to Europe through Turkey, and the Turkey – Greek – Italy Gas Pipeline (SEGR) are both projects that are being executed by cooperation among several countries and financially supported by the European Union.

Another important aspect for the EU is its future bilateral negotiations and agreements with traditional suppliers to ensure a balance. Russia is one of the most important traditional suppliers to the EU and is trying to keep its position in the future. However, Russia may not be able to adequately satisfy the growing demand on both its export and domestic markets[8]. Over the past decade, EU-Russian relations have encompassed a broad set of policy issues. Among them, the issue of energy has played a crucial role. Indeed, Russian oil and gas exports are achieving ever-increasing importance as part of the country's foreign policy[9]. The relationship between the EU and Russia in terms of balancing expectations and interests is as follows: Russia wants to secure the energy demand presented by the EU market and to gain a stronger foothold in the EU internal market, while the EU seeks non-discriminatory and fair treatment from Russia in their energy relationship[10].

2. Natural Gas Policy of Turkey

Natural gas reserves are very limited in Turkey, and the economic profitability of production is low. By the end of 2005, natural gas production totalled 27 billion m^3, while consumption totalled 800 billion m^3. Domestic production accounts for just 3% of total consumption, because reserves and production are very limited, and consumption has rapidly increased since 1985, and grew by 15% on average between 1989 and 2005. Turkey has been the seventh biggest producer since 2005 and accounts for 5% of total European consumption.

7 Dikkaya, Mehmet, (2009), Nabucco Projesi: Tarihin Yeniden Canlanişi mi?, http://www.mehmetdikkaya.org/images/stories/GUNCEL_YAZILAR/Nabucco_Projesi.pdf.

8 Communication From the Commission To The European Council 'External energy relations: From principles to action' Com (2006) 560 final, Brussels, 2006, p. 4.

9 European Commission, (December 17, 2002) 'EU–Russia energy dialogue – Progress Report', published in Energy Europe, pp. 1-16.

10 Communication From the Commission To The European Council 'External energy relations: From principles to action' Com (2006) 560 final, Brussels, 2006, p. 4.

The USA and European Union see Turkey as a *strategic cooperation partner* for providing energy security. Equally, as Turkey is also dependent on Russia for 70% of its energy, it is just as interested in securing energy supply and diversifying energy resources as the European Union and the USA. Accordingly, common pipeline projects and substructure works are aimed at achieving these goals.

Resource dependence on just one source of natural gas harbours relatively more risk for Turkey than petrol. As mentioned before, the structure of natural gas transfer is less elastic than petrol. In this respect, Turkey must diversify its sources of natural gas in case supply from Russia is cut. In this context, the country is striving to diversify its natural gas resources within the Caucasian, Middle Asia and Khazar Regions, which are neighbouring, or culturally close. Examples of these efforts include the gas purchasing agreements signed in the 2000s, especially with Iran and Azerbaijan.

Turkey's gas demand is increasing and cannot be met by Russia alone.

Traditionally, Russia has been Turkey's largest gas supplier. When we look at the current economic condition of Russia, the Caspian energy resources take on a much more important role for Turkey's energy demand. The Caspian energy resources could therefore be an important alternative for Turkey in terms of its own consumption and transit. The areas of supply and demand need to be directly connected by pipelines, as shipping gas is technically very complex and expensive[11]. The EU and Turkey therefore joined forces in a project that was financially supported by both Turkey, related countries along the pipelines from the Caspian region to Europe, and the EU.

Turkey has continued its efforts to diversify resources and routes to strengthen its role as a transit country for oil and gas from the Caspian Basin and the Middle East to the EU. The main goals of Turkey's energy policy:

- To prioritize consumption of domestic energy resources to meet demand
- To diversify energy sources and to avoid dependence on energy imports,
- To be a transit country by transporting energy sources produced in the Caspian Region to the European markets,
- To contribute to the realization of the East West Energy Corridor and to improve natural gas supply diversification and security both for Europe and itself[12].

11 European Commission: Turkey 2005 Progress Report; COM (2005) 561 final; http://europa.eu.int/comm/enlargement/report_2005/pdf/package/sec_1426_final_en_progress_report_tr.PDF.

12 www.botas.gov.tr.

3. The EU's Natural Gas Policy and Security

The most important basic argument in energy security policy is the geopolitical work needed to access the natural resources. Work must focus on various potential energy resources and providing diversified supply (petrol, gas, coal, nuclear energy and renewable energy). Countries' energy policies also offer a basis for studying the relations between economic development and growth[13].

The European Commission has defined *energy security* as "meeting the expected energy needs with national resources, continuing strategic reserves under economically suitable conditions, obtaining energy from suitable strategic stocks, and the accessing energy from trustworthy foreign resources".

Following the economic modernization of the world in the 1980s, global energy demand has increased. The European Union's demand for energy also continues to grow. It is obvious that the EU is in a process of enlargement and deeper economic integration. The role of energy policy on the EU's current agenda is becoming a major internal and external issue.[14].

Europe accounts for17% of total world gas consumption[15]. Energy consumption is generally increasing in most of the regions around the world, and especially in developing countries. The EU, however, is not the only part of the world with high energy demand: India and China consume significant volumes of energy with their growing populations and ongoing process of economic modernization. Therefore, Europe is not the only region where growth results in additional energy demand. The question facing the EU is that of a sustainable energy supply, since the growth in China and India has also caused a significant amount of new energy demand. Even in the energy rich region of the Middle East, population growth coupled with economic modernization has resulted in higher energy demand. Other parts of the world, such as Africa and Latin America, are also improving their standard of living and economies[16]. Consequently, all of these are signs of increasing global energy demand.

---

13 Proops, J, (1998), Energy Intensities, Input–Output Analysis And Economic Development. In M. Giaschini ed. Input–Output Analysis: Current Developments, New York and London: Chapman and Hall, p. 210.

14 Aad, C., Coby L. (2006) 'Energy supply security and geopolitics: A European perspective' Elsevier Energy Policy, p. 532-543.

15 Vince L. Morelli Analyst in European Affairs Foreign Affairs, Defence and Trade Division, CRS Report for Congress: The European Union's Energy Security challenges (Report. Congress), 9.11.2006, p. 11.

16 Vince L. Morelli Analyst in European Affairs Foreign Affairs, Defence and Trade Division, CRS Report for Congress: The European Union's Energy Security challenges (Report. Congress) 9.11.2006, p. 11.

On the other hand, the Ukrainian gas crisis, which broke out in 2006, undermined for EU policies. As Russia cut off supply in the first days of the year, the EU came within an inch of an energy crisis. This situation, which demonstrates how energy can be used as an instrument of policy when necessary, has strengthened the EU's resolve to implement a common energy policy[17] to mitigate the risk associated with unilateral and strong dependence on Russia. The issue of border security in connection with the supply of gas to the EU through the Ukraine also became a matter of argument in 2006[18].

The European Union's energy system is based on three underlying factors: energy security, ensuring competition in the energy sector, and environmental protection. It is thought that energy exports from the European Union will increase at a stable rate from 2010 to 2020. [19]. The EU, as a whole, doesn't have enough energy resources to meet its own energy consumption. Thus, it must meet its demand for fossil energy from countries outside the Union. In this context, European Union energy policy instruments address the issues of controlling demand, differentiating European resources, making domestic energy markets more effective, and controlling foreign supply. To control foreign supply, the European Union aims to strengthen alliances with producer countries and regions, for which it is collaborating with Russia, Northern Europe, the Baltic region and North Africa in terms of energy. Basically, the recent position of the European Union has been that increasing consumption requires new energy routes for EU members[20].

The EU initially did not have a common energy policy as the members were all applying different energy policies, until the European Council endorsed its Green Paper in March 2006. This paper aimed to ensure long-term sustainable energy supply and competitiveness at an international level. The EU is trying to strengthen economic competitiveness, increase diversity, improve efficiency, and foster investments and innovation.

The 27 member states of the European Union have transferred a certain level of their sovereignty (or competencies) to the EU institutional structure. But, energy policy for the most part remains primarily the responsibility of the member

---

17 *Worsening Geopolitics and Increasing Global Competition for LNG*, Oxford Institute for Energy Studies, October (2006), p. 2.

18 Gnedina Elena Gnedina and Emerson Michael, (2009), The Case for a Gas Transit Consortium in Ukraine: A Cost-Benefit Analysis, *Centre for European Policy Studies*, No. 180, January 2009, p. 1.

19 EUROPEAN COMMISSION, 2000, *Towards a European Energy Strategy for the Security of Energy Supply*, European Commission Com 2000-769 Final.

20 BACIK, Gökhan, 2006, Turkey and Pipeline Politics, *Turkish Studies*, Vol. 7, No. 2, 293-306, June 2006, p. 295.

states[21]. However, implementing the EU's energy policy is the function of the European Commission that has been able to exercise a significant amount of influence over energy policy through its authority to pursue internal market competition, environment and consumer protection policies[22].

The EU energy market is growing increasingly dependent on imports. Europe's main gas corridors today run either through or from North Africa and connect to the EU grids through Italy and Spain; or, from Russia though the Ukraine to Central and Western Europe. Turkey's geopolitical position is significantly important; its closeness to the most important gas field of central Asia, the Persian Gulf, Iran and Russia has made Turkey the most attractive gateway for the 'third artery' of the EU' s energy supply[23].

4. Role of Russia

Russia has significant energy reserves within its borders, and is traditionally the most important supplier to the EU. Russia may, however, not be able to meet the growing demand from both its export and domestic markets adequately because of its inadequate investment in the production, transportation and distribution of energy. Russia is trying to keep its position in the future for Europe. EU-Russian cooperation is therefore significant for the security of the EU, because Russia is already the origin of around 25% of the oil and gas consumed in the EU[24]. All natural gas is exported GASPROM, which holds a monopoly in Russia[25]. This monopolistic position of GASPROM makes it difficult for member countries to negotiate purchasing agreements with GASPROM since GASPROM can choose if it wants to enter into an agreement or not. On the other hand, Khazar, the Middle Asian countries and the EU are pursuing new projects to decrease their dependence on Russia. However, the present political structure shows that it would be difficult to create an energy alliance that were totally contrary to Russia's

21 Vince L. Morelli Analyst in European Affairs Foreign Affairs, Defence and Trade Division, CRS Report for Congress: The European Union's Energy Security challenges (Report. Congress), 9.11.2006, p. 3.

22 Vince L. Morelli Analyst in European Affairs Foreign Affairs, Defence and Trade Division, CRS Report for Congress: The European Union's Energy Security challenges (Report. Congress), 9.11.2006, p. 4.

23 Walter Posch and Barut Grcic 'Strategic Implications for Central Europe' Policy Analyzing, Ljubljana Institute For Strategic Studies, December 2004

24 EU Commission Report: External energy relations: From principles to action (Com. Report. 560 final), Brussels, 12.10.2006, p. 4.

25 *Worsening Geopolitics and Increasing Global Competition for LNG*, Oxford Institute for Energy Studies, October 2006, p. 3.

benefits, and the contributions of Russian enterprises will be required for such an alliance[26]. Energy is one of the items that have been the focus of intense negotiations between the EU and Russia over the last ten years.

After the 1990s, the European Union's energy policy changed, with basic emphasis shifting to *diversifying energy supply, competition, and environmental protection*[27]. It has been reshaping its relationship with Russia, which holds 13% of the world's oil and 32% of all gas reserves. The European Union and Russia have tried to develop energy dialogues to support long term energy cooperation, and this has made it compulsory for the European Union to support Russian energy substructure projects that are necessary to secure the supply of oil and gas. The EU believes that a trustworthy cooperation framework should be created for developing trade with Russia, and encourages Russia to take part in multilateral international agreements such as the Energy Charter Treaty and Kyoto protocol.[28] The relationship between Russia and the European Union has strategic dimensions for both sides. Although both sides have their own benefits, the position occupied by Russia in EU calculations must be observed in three dimensions in terms of energy security policy.

- The *political dimension* related to decreasing dependence on energy imports;
- The *economic dimension* named as creating a competitive market and providing energy price variety.
- The dimension for providing international decisions that play a role as an additional systematic factor relating to energy policies.

Since 2000, the EU and Russia have been improving the Energy Dialogue, which aims to secure the long-term mutual energy partnership. This political agreement includes a commitment by the EU to support those Russian energy infrastructure projects that are necessary for supplying oil and gas to Europe. It is necessary to continue a stable pattern for cooperation, and to encourage Russia to actively participate in major international multilateral frameworks (institutions) such as the Energy Charter Treaty and the Kyoto Protocol[29].The mutual expectations and benefits of Russia and the EU can be summarized as follows:

---

26 *Kassenova, Nargis, (2009), Kazakhstan and the South Caucasus corridor in the wake of the Georgia-Russia War, EUCAM EU-CENTRAL ASIAMONITORING, No. 3 – January 2009.*

27 EUROPEAN COMMISSION, (1996), *White Paper For Energy Policy*, COM 95 682, http://europa.eu/documents/comm/white_papers/pdf/com96_421_en.pdf.

28 EUROPEAN COMMISSION, (2002, 2001), Annual Energy Review, Synthesis Report on EU–Russia Energy Dialogue, October 2001.

29 European Commission, December 17, 2002) 'EU–Russia energy dialogue – Progress Report', published in Energy Europe, pp. V 1-16.

– Russia is trying to keep its position as the biggest supplier to the EU
– The EU wants to ensure fair treatment for sustainable energy cooperation[30].

This process has accelerated the EU debate on energy security and ensured that it is a priority issue. In January 2006, when assessing the impact of the Ukraine gas crisis, EU Energy Commissioner Piebalgs reiterated that Europe needed "a more cohesive policy on the security of its energy supply[31].

"Russia not only wants to meet the energy demand of EU but also wants to meet a good portion of African and United States demand. Energy is an important political force for Russia and it is also a major factor for its economic development. For some, it is acknowledged that the Putin government has made it clear that it intends to use its energy export power to regain Russia's cold war influence around the world[32]."

There are existing and future bilateral agreements with energy producer and transit countries such as the PCAs (Partnership and Cooperation Agreement), the Memorandum of Understanding on Energy Cooperation with the Ukraine and the Association Agreements with Mediterranean countries. In addition, there are the European Neighbourhood Policy Action Plans; the foreseen memoranda of understanding with Algeria, the Azerbaijan and Kazakhstan Euromed energy cooperation, and the EC-Norway energy dialogue[33].

Turkey is becoming a crucial energy corridor for supplies from the producer regions, which makes Turkey strategically important for the EU's energy security. The enlargement process with Turkey could contribute to promoting the early adoption and implementation of the EU's energy acquis by Turkey, while Turkey's early accession to the Energy Community Treaty could also speed up this process[34]. The Energy Community Treaty came into force on 1 July 2006 and extends the relevant EU energy acquis to the Western Balkan countries. The implementation of the Treaty will improve energy security, create a regional energy market and encourage vital investments[35].

30 EU Commission Report: External energy relations: From principles to action (Com. Report. 560 final), Brussels, 12.10.2006, p. 5.

31 "Lessons for EU from Gas Crisis", BBC News, January 4, 2006.

32 Keith Smith, Center for Strategic and International Studies, testimony before the Congress: Russian Energy Policy and its Challenges to Western Policy Affairs (Congress. Report.), 5.16.2006, p. 7.

33 EU Commission Report: External energy relations: From principles to action (Com. Report. 560 final), Brussels, 12.10.2006. p. 5.

34 EU Commission Report: External energy relations: From principles to action (Com. Report. 560 final), Brussels, 12.10.2006. p. 5.

35 EU Commission Report: External energy relations: From principles to action (Com. Report. 560 final), Brussels, 12.10.2006, p. 5.

5. The Nabucco and South European Gas Ring (SEGR) 'Turkey-Greece–Italy Gas Interconnector' Projects

Providing supply security ranks very high up on the European Union's energy policy targets. In an attempt to become less dependent on the Middle East and Russia for providing natural gas and oil supply security, it is focusing more on the resources in the Caucasus and Khazar regions, and joint regional energy ventures in the Mediterranean and Black Sea as a means of providing regional variety. An evaluation of the relationship between Turkey and the European Union from an energy perspective reveals the strategic importance of Turkey for providing energy supply security to the EU[36].

It is time for the EU to take action, as it still hasn't agreed a common energy policy, especially after the last Ukrainian Crisis. All of the gas routes to the EU still pass through Russia. Accordingly, the EU obtains 80% of its gas needs through the Ukraine. Its search for alternative routes has prompted the EU to create and support new projects. NABUCCO, a name that was derived from Nebuchadnezzar, the King of Babylon, is found in the Torah, and was made famous as the title of an opera by Italian composer, has now been brought from the dusty pages of history to the corridors of EU institutions and to the minds of the diplomats of the countries related to the project[37].

The project has been put forward for transferring combined Caucasian and Middle East natural gas to Europe through Turkey. The project, which aims to transfer nearly 30 billion cubic metres of natural gas to Europe annually, was launched in 2002, and moved forward by a consortium constituted under the leadership of Austria. During the pipeline construction period, at least 5,000 people are expected to be employed, while the investment amount is estimated to be 4.5-5 billion Euro. A huge proportion of the project will be financed by the EU[38].

The length of the pipeline passing through each related country will form the calculation basis for distributing the tax income generated from the pipeline by the agreement. Accordingly, 60% – the largest share of the tax income – will go to Turkey when the pipeline commences operation. As a result, Turkey is expected to earn income of 400-450 million Euro per year.

The point under discussion is whether the NABUCCO project will transfer enough gas to Europe. The gas resources in Khazar and the Middle East coun-

---

36 Oktay, Ertan, Çamkiran, RADİYE, Funda, 2006, Avrupa Birliği'nin Güvenliği Açisindan Türkiyenin Önemi, *Avrupa Araştirmalari Dergisi*, Cilt 14, Sayi 1, p. 17.

37 Dikkaya, Mehmet, 2009, Nabucco Projesi: Tarihin Yeniden Canlanişi mi?, http://www.mehmet dikkaya.org/images/stories/GUNCEL_YAZILAR/Nabucco_Projesi.pdf.

38 The Business, 2007.

tries – proven to amount to 83 trillion cubic metres – that could be supplied to the European gas market are thought to be much more than the total capacity of the Turkey – Greek natural gas pipeline and Nabucco Projects[39]. The NABUCCO project is comprises a series of natural gas pipeline systems, and aims to transport some of the huge gas reserves from the Khazar region and Middle East to Europe through Turkey. This pipeline crosses the borders of five countries: Bulgaria, Romania, Hungary and Austria, in addition to Turkey[40].The "Turkey-Greek-Italy Pipeline Project" is an alternative project pursued by the European Union, which aims to pipe the reserves in the Middle East and Khazar region to Europe through Greece, Italy and Turkey. The project is expected to pipe 31 billion cubic metres of natural gas to Europe each year[41].

According to BP's statistics of 2006, the amount of natural gas supplied to Europe by the NABUCCO project will be more than the total supplied through the other energy routes (Russia, North Africa, North Sea, and own production) in 2030. Accordingly, it is estimated that the NABUCCO project will be the world's highest-capacity natural gas pipeline by 2030.

6. Conclusions

Energy is a significant factor for countries. Moreover, the concept of security of supply is a common agenda for countries and the EU. Russia, a self-supporting and net exporting country, has more opportunities than other economies in terms of economic growth and development thanks to its resources. When it comes to politics, Russia never hesitates to use energy as a policy instrument whenever necessary. Russia is the most important energy supplier to the European Union and is striving to sustain this position in the future, too. For this reason, cooperation between the European Union and Russia is extremely important to ensure the EU's energy security. On the other hand, sustainable energy exports play a hugely important role for the Russian economy. It is energy exports that are driving economic development in Russia.

Another important factor that affected the EU's energy policy was the Ukraine gas crisis, which raised a number of critical policy questions and set in

39 TÜSİAD, (2007), Avrupa Birliğine Katilim Sürecinde Türkiye'nin Komşu ve Çevre Ülkeler Politikasi, Yayin No. TÜSİAD-T/2007-05/434, p. 312.

40 UNECE Gas Center, December 2005, *Report on The Gas Company and Market Structure in The European Part of The Unece Region*, Geneva, p. 17.

41 İpek, Pinar, 2006, The Aftermath of Baku-Tbilisi-Ceyhan Pipeline: Challenges Ahead for Turkey, *Perceptions*, Spring 2006, p. 5.

motion a series of events that moved energy security to the forefront of Europe's anticipated energy policy debate. The last crisis prompted the European Union to search for alternative energy routes. This search for alternatives to Russia has empowered the dimension of integration among the member states, and has accelerated the process of integration with the transit countries who supply energy during the cooperation process. Turkey plays a key role as a transit country, given that it borders on the world's most energy-rich regions and as a means of providing energy supply to Europe. Since Azerbaijan, Turkmenistan and Kazakhstan do not have sea access, they must cooperate with their neighbouring countries to bring their energy resources to the world markets. Turkeys has the best geopolitical position in this respect. The countries who want to lessen their energy dependence on Russia are enthusiastic about cooperating with Turkey. Turkey doubtlessly offers the best alternative route, thanks to both its geopolitical role and its candidate status for accession to the European Union. Because of Turkey's position, the length of pipeline planned to pass through Turkish territory has meanwhile been extended.

In order to be successful in terms of sufficiently regulated natural gas markets that meet European standards, energy cooperation with the EU is a key factor for Turkey's ultimate accession to the EU. Although, member states cite economic and political disadvantages of Turkey joining the EU, the strategic importance of Turkey in satisfying the EU's demand for natural gas and the successful implementation of these projects will show how necessary future Turkish membership to Europe is in terms of their security.

References

Aid C., Cobi L., (2006) 'Energy supply security and geopolitics: A European perspective' Elsevier Energy Policy.

Aldo S. (2006) 'Russian gas price reform and EU-Russia gas relationship: Incentives, consequences and European security of supply' Elsevier Energy Policy, November.

Andreas G. (2007) 'Rhetoric reality: Russia threats to European energy supply' Elsevier Energy Policy.

Andrei B. (2003) 'New Dimension of Energy Security of The Enlarging and Their Impact on Relations With Russia' European Integration, December.

Bacik, Gökhan, 2006, Turkey and Pipeline Politics, *Turkish Studies*, Vol. 7, No. 2, 293-306, June 2006.

BBC News, "Lessons for EU from Gas Crisis" January 4, 2006.

Demirbas Ali, (2000), Recent advances in biomass conversion technologies, Energy Ed Science and Technology, Vol. 6.
ETUC Executive Committee Report: The European Energy Policy (Report Committee), Brussels, 14-15, March 2006.
EU Commission Report: External energy relations: From principles to action (Com. Report. 560 final), Brussels, 12.10.2006
European Commission, 2000, *Towards a European Energy Strategy for the Security of Energy Supply*, European Commission Com 2000-769 Final.
European Commission, 1996, *White Paper For Energy Policy*, COM 95 682, http://europa.eu/documents/comm/white_papers/pdf/com96_421_en.pdf.
European Commission, 2002, 2001, *Annual Energy Review, Synthesis Report on EU – Russia Energy Dialogue, October 2001*, http://www.uni–mannheim.de/edz/pdf/sek/2002/sek–2002-0333–en.pdf.
European Commission (December 17, 2002) 'EU – Russia energy dialogue – Progress Report', published in Energy Europe.
European Commission: Turkey 2005 Progress Report; COM (2005) 561 final; http://europa.eu.int/comm/enlargement/report_2005/pdf/package/sec_1426_final_en_progress_report_tr.pdf.
İpek, Pinar, 2006, The Aftermath of Baku-Tbilisi-Ceyhan Pipeline: Challenges Ahead for Turkey, *Perceptions*, Spring 2006.
Gnedina, Elena Gnedina and Emerson Michael, 2009, The Case for a Gas Transit Consortium in Ukraine: A Cost-Benefit Analysis, *Centre for European Policy Studies*, No. 180, January 2009.
International Energy Agency, 2004, *Security of Gas Supply in Open Markets*, http://www.iea.org/textbase/nppdf/free/2004/security.pdf.
Kassenova, Nargis, 2009, *Kazakhstan and the South Caucasus corridor in the wake of the Georgia-Russia War, EUCAM EU-CENTRAL ASIAMONITORING, No. 3 – January 2009.*
Kiliç Ali (2005), Turkey's natural gas necessity, consumption and future perspectives, Energy policy, Vol. 1.
Morelli V. L. (2006), Analyst in European Affairs Foreign Affairs, Defense and Trade Division, CRS Report for Congress: The European Union's Energy Security challenges (Report. Congress), November 9.
Oktay, Ertan, Çamkiran, RADİYE, Funda, 2006, Avrupa Birliği'nin Güvenliği Açisindan Türkiyenin Önemi, *Avrupa Araştirmalari Dergisi*, Cilt 14, Sayi 1.
Smith K. (2006), Center for Strategic and International Studies, testimony before Congress: Russian Energy Policy and its Challenges to Western Policy Affairs (Congress. Report.), May 16.

Proops, J, 1998, *Energy Intensities, Input–Output Analysis And Economic Development.* In M. GIASCHINI ed. *Input–Output Analysis: Current Developments*, New York and London: Chapman and Hall.
TPAO Report: Annual Activation Report (Annual Activation Report), Ankara 2002
TÜSİAD (2002) Regulatory Institutions and Application of Turkey, (TÜS; AD Publishing, No.-T/2002-12/349, December 2002).
TÜSİAD, 2007, Avrupa Birliğine Katilim Sürecinde Türkiye'nin Komşu ve Çevre Ülkeler Politikasi, Yayin No. TÜSİAD-T/2007-05/434
Unece Gas Centre Report: Gas Company and Market Structure in the European Part of the Unece Region (Co. Report.), Geneva, December 2005.
UNECE-Sixteenth Session of the Working Party Report: Natural Gas Projects In Turkey Recent Developments (COM. Report.), 24-25.1.2006
Walter P. , Barut G. (2004), Strategic Implications for Central Europe, Policy Analyzing, Ljubljana Institute For Strategic Studies, December.
Worsening Geopolitics and Increasing Global Competition for LNG, Oxford Institute for Energy Studies, October 2006.
www.axisglobe.com/article.asp?article=311.
www.botas.gov.tr/eng/projects/bulgaria.asp.
http://ec.europa.eu/dgs/energy_transport/international/regional/euromed/energy/doc/conference_4/2003_11_20_annexes_en.pdf.
www.epdk.org.tr/yayin_rapor/yillik/yillik.htm.
www.sofiaecho.com/article/pipeline-projects0back -track/id 9599/catig 23

Future of Russian Gas and its Energy Relations with the EU between now and 2020

Yuri Yegorov and Franz Wirl

1. Introduction

The market for natural gas is a very dynamic industrial sector. Recent studies have shown the important role of geography, politics and technology for its evolution. The gas market differs from other markets due to its very high share of transport and infrastructure costs. Since investment is location specific, it involves geopolitical aspects.

Russia and the EU are mutually dependent on energy relations and maintain a kind of monopolistic-monopsonistic relationship. Ericson[1] analyses the Eurasian network of natural gas pipelines. The EU is concerned about its energy dependence on Russia. The technology and high infrastructure costs make gas markets highly inflexible. Very often economic considerations take second place to political issues. Astronomical up-front investment costs lead to the emergence of natural monopolies. The development of such a system requires long-term contractual relationships that tie producers and consumers together. Contrary to oil, natural gas is typically supplied through a singular gas pipeline network. The pipeline network can impose "lock-in" relationships between a producer and consumers. The EU dependence on Russian gas is about 30%. Gazprom receives 65% of its revenues from Europe, and generated 8% of Russian GDP in 2005. 40% of the Russian budget depends on the export of energy. Thus, a mutual dependence between buyers and sellers exists, and this raises security issues for both sides of this tied market.

At the same time, EU policy towards Russia at present does not allow a mutually beneficial outcome. On the one hand, Europe correctly understands the necessity of security of gas supply. But it plans to do it at the expense of Russia and thus puts Russia in a difficult position. The policy to diversify gas supplies (and thus to reduce the present share of 30% of Russian gas in EU consumption) might be short sighted, as has been shown by Yegorov and Wirl.[2] The reason is that the gas reserves of alternative suppliers are only sufficient for a middle term

1 Ericson R. (2009) Eurasian Natural Gas Pipelines: The Political Economy of Network Interdependence. – *Eurasian Geography and Economics*, Vol. 50, No. 1, p. 28-57.

2 Yegorov Y., Wirl F. (2008) Energy relations between Russia and EU with emphasis on natural gas, *OPEC Energy Review*, 2008, Vol. 32, iss. 4, p. 301-322.

horizon, and in the long run Europe will be more dependent on Russia, which meanwhile can enter into alternative commitments (such as gas pipelines to China).

In his paper, Finon[3] criticises that the EU's energy policy confronts Russia's interests. Departing from different economic theories, he shows the advantage of South Stream versus Nabucco, and the advantage of long-term commitment between producers and consumers for the development of infrastructure. The total cost advantage of Nabucco versus South Stream is not obvious. Besides, Russia has an advantage in long-term contracts. Downstream, Nabucco has an advantage over South Stream: its shareholders are historic companies of transit countries. However, South Stream has an upstream advantage in Gazprom's possibility to control multiple sources of supply.

This article attempts to put on "Russian shoes" and see what rational strategies it might have for future gas development, depending on the evolution of its political relationships with Europe, the dynamics of world prices for natural gas, and the level of competition for gas from Central Asia.

Typical literature about natural gas is either fully descriptive, with detailed case studies (like the books edited by Pirani[4] and Victor, Jaffe and Hayes[5]) or presents some relatively complex model[6] or numerical results from a theoretical model[7]. This is a kind of analytical essay that combines elements of descriptive analytics with simple microeconomic reasoning. While the object of study (the influence of geography and politics on the economics of gas markets) is similar, it nevertheless differs from other studies of gas markets by Yegorov and Wirl. Yegorov and Wirl[8] have analysed static transit games between Russia and Turkmenistan, whereas here, the focus is on optimization of Russia in the environment of heterogeneous costs, Russian incentives for particular investment and the stylized scenario of the development of gas markets involving Russia in the middle term horizon (next decade).

---

3 Finon D. (2009) The limits of the EU direct foreign gas policy. Autopsy of the stillborn Southern corridor project Nabucco, CIRED WP, December 2009, 17 p.

4 Pirani S., Ed. (2009) Russian and CIS Gas Markets and Their Implications on Europe, Oxford University Press.

5 Victor, D.; Jaffe, A.; Hayes, M., Eds. (2006) *Natural Gas and Geopolitics: From 1970 to 2040*, Cambridge University Press, 488 p.

6 Hubert F., Ikonnikova S. (2004) Hold-Up, Multilateral Bargaining, and Strategic Investment: The Eurasian Supply Chain for Natural Gas. – WP, Humboldt University, Berlin.

7 Hartley p. , Medlock K. (2009) Potential Futures for Russian Natural Gas Export. – *The Energy Journal*, Special Issue. World Natural Gas Market and Trade: A Multi-modelling perspective, p. 73-95.

8 Yegorov Y., Wirl F. (2010a) Gas Transit, Geopolitics and Emergence of Games with Application to CIS Countries, USAEE – IAEE WP No. 10-044, February 2010, 29 p. (http://ssrn.com/abstract =1560563).

Section 2 discusses why the fast development of new Russian fields in the North may not be an economically profitable strategy under present conditions. Section 3 describes gas resources of Turkmenistan. Section 4 is devoted to the problem of gas transit. Section 5 presents theoretical model. Section 6 is about policy implications. Section 7 concludes

2. Costly Gas in Russian North

Russia has 26% of the world's proven reserves of natural gas. This is the leading position in the world and an even higher concentration than exists for the oil market. For political and technological reasons, the two countries that follow Russia – Iran (15%) and Qatar (14%) – are still not exploiting their full potential. In gas supply to Europe, Russia mainly competes against Norway and Algeria, who each have less than 3% of the world's reserves. Thus, if Europe chooses more competition in the short run, it will face less competitive gas markets in the long run, after the reserves of smaller competitors have dried up.[9] However, Russia's major problem is the high cost of gas extraction in new fields (Yamal, Shtockman) that are located in the North and Arctic shelf. The gas fields that are used today are in natural decline and Russia needs to invest a lot in developing new fields[10]. It is important to note that most of the gas pipeline infrastructure was built in Soviet times, when the costs were much cheaper than today. There are several reasons why the necessary investment in Russia is too costly at present: severe climate conditions, the strong national currency, and corruption. Western investors might not come in when they recall what happened with Sakhalin. Russian companies have never had experience of investing hundreds of billion dollars (and this is the scale of required investment), and even powerful Gazprom is trying to get the state to subsidize the development of Northern fields.

Investment theory teaches us that each project has a net present value (NPV). If p_t denotes prices in period t, Q_t denotes volumes, c_t denotes variable cost, F – fixed cost, T is time horizon and r is discount (interest) rate, then the formula can be written as follows:

$$NPV = \Sigma^{T}_{t=1} (p_t - c_t)Q_t(1+r)^{-t} - F.$$

9 Yegorov Y., Wirl F. (2008) Energy relations between Russia and EU with emphasis on natural gas, *OPEC Energy Review*, 2008, v.32, iss.4, p. 301-322

10 Stern J. (2009) Future Gas Production in Russia: is the concern about lack of investment justified? – Oxford Institute for Energy Studies, NG 35, 17p.

Consider, for simplicity, a constant price, p, that is expected by investors. Then it is easy to show that $\partial(NPV)/\partial p>0$, while $\partial(NPV)/\partial r<0$. The interest rate in Russia was always high compared to Europe, which puts it at a disadvantage in terms of investment. Since these gas development projects in the North involve very high F (value of fixed investments up front), a substantially high expected price p is a pre-condition of economic viability (NPV>0) of the investment project. And in the present environment, Russia cannot be sure it will secure a substantially high price (given the gas price stagnation over the last 2 years), nor does it have any certainty about the volume Q, which Europe will need (especially now that the latter's is moving away from long term contracts with Russia, with the gas price linked to the oil price).

From the perspective of a central planner, if a country has deposits of resources with heterogeneous extraction costs, it makes more sense to first develop the cheaper resource, and then to move on to the more expensive. In the days of the USSR, the gas resources in the South (where production is cheaper, not least because infrastructure investments are cheaper) were exploited first, while the more costly resources in the North were frozen. After the USSR was dissolved, Russia could no longer control gas extraction in Central Asia, as it no longer had any property rights[11] but instead can only extract transit rents.

3. Natural Gas in Turkmenistan

The natural gas resources in the South of the USSR were already important before its dissolution. Among the independent countries of the region, Turkmenistan plays a key role as a producer, exporter and holder of natural gas reserves. At present, Turkmenistan is the second largest gas exporter from the CIS. When it gained independence in 1991, Turkmenistan was the 4th largest producer of gas in the world, after Russia, the USA and Canada. Its historical peak of production was reached in Soviet times, at 89.9 bcm in 1989.[12] Later, its gas production and exports went down, to a historical low in 1998, but had recovered by 80% by 2007 when it produced 72.2 bcm, of which 51 bcm were exported.

The story of Turkmenistan is quite interesting; it is very important and is surrounded by a lot of uncertainties. The detailed description of its gas fields is pro-

11 Olcott M. (2009) Russia, Central Asia and the Caspian: How Important is the Energy and Security Trade-off? – WP, James Baker III Institute for Public Policy of Rice University, USA, 31 p.

12 Pirani S., Ed. (2009) Russian and CIS Gas Markets and Their Implications on Europe, Oxford University Press.

vided in[13]. There is still huge uncertainty regarding its gas reserves. BP estimates the lower limit at 2.67 tcm.[14] Turkmen authorities claim 4.9 tcm of proven reserves and 24.6 tcm of potential reserves, but this is not yet supported. Still, even BP estimates put Turkmenistan close to Norway in terms of reserves, and Norway is the second largest gas exporter to the EU, covering 18% of EU gas demand.

The domestic consumption of gas ranges from 8-15 bcm/yr. About 35% goes to electricity generation, 55% to industrial consumers and 10% to households. Although former president Niyazov decreed in 1992 that gas and electricity be supplied free to residents, and kept industrial prices for gas much below international levels, it did not cause any problems for the country's revenues. The lion's share of extracted gas is exported; gas is the main export product (accounting for 55% of total exports in 2007) and these gas exports drive the country's GDP. The physical volumes of exported gas ranged from just 1.7 bcm in 1998 to 51.2 bcm in 2007. The export price also varied extensively, from $35/tcm in 2000 to $96/tcm in 2007, with a tendency for further growth.[15]

Turkmenistan is a good example of a country where geography and politics, rather than economics, drive the development of gas markets. Here, it is not production that is the major issue, but market access. In USSR times, Turkmenistan was well connected to the gas markets via pipelines that passed through the Russian Federation and other Soviet republics in Central Asia. After the dissolution of the USSR the physical structure remained, but the legal environment has changed, and Russia now occupies a monopolistic position in terms of the transit of Turkmen gas. All attempts to overcome this situation have not been successful. Olcott[16] describes the long history of a plan to build a gas pipeline to Pakistan and India via Afghanistan. Although this project got political support from the USA, the plan was never implemented for security reasons, as it was felt to pose an unacceptable risk for investors. Equally, a pipeline via Iran would be feasible, but the embargo of this country means that this project is also blocked. The Trans-Caspian pipeline link between Turkmenistan and Azerbaijan is in the interest of European investors (since it can provide the required capacity to Nabucco), but this project is unlikely to be implemented in the near future, due to the strong opposition from Russia and Iran based on environmental concerns. Such situations show that many investment projects for natural gas are "frozen"

---

13 Pirani S., Ed. (2009) Russian and CIS Gas Markets and Their Implications on Europe, Oxford University Press.

14 BP Statistical Review (2008).

15 Pirani S., Ed. (2009) Russian and CIS Gas Markets and Their Implications on Europe, Oxford University Press.

16 Olcott M.B. (2006) International gas trade in Central Asia: Turkmenistan, Iran, Russia and Afghanistan. In: *Natural Gas and Geopolitics: From 1970 to 2040*, D. Victor, A. Jaffe, M. Hayes, Eds.; Cambridge University Press, pp. 202-233.

because of "bad geography" and political reasons. Later this argument will be applied to the "frozen" scenario for the next decade.

4. Transit of Gas from the Caspian Region

Turkmenistan is a key country for gas in the Caspian region. Table 1 shows gas reserves, production and export for 4 countries in the area. In 2008, new reserves were added by BP to Turkmenistan, putting it into 4th position in the world (after Russia, Iran and Qatar). Although it produces gas at quite a modest speed (at its present production level, reserves are sufficient for 120 years), almost on a par with Uzbekistan, its low share of domestic consumption means that it dominates the Caspian region in terms of gas exports. That is why strategic considerations concerning the Caspian area focus on Turkmenistan.

Table 1: Natural gas reserves, production, consumption and export by Caspian countries

Country	Reserves, tcm	Production, bcm (2008)	% world reserves	R/P rat., years	Consumption, bcm	Export*, bcm
Azerbaijan	1.2	14.7	0.6	81.6	9.3	5.4
Kazakhstan	1.82	30.2	1	60.3	20.6	9.6
Uzbekistan	1.58	62.2	0.9	25.4	48.7	13.5
Turkmenistan	7.94	66.1	4.3	120.1	19	47.1

Source: BP Statistical Review 2009, calculation by authors (R/P, export)

The geographical position of these four countries differs. While Azerbaijan has direct access to the European market (via Georgia and Turkey), its reserves are the lowest among these 4 countries, and its strategic role is correspondingly less important. Kazakhstan has direct access to the Chinese market, and is also strategically important in terms of gas transit. Only Uzbekistan and Turkmenistan have no direct access to gas markets and thus require transit cooperation. Since Uzbekistan has substantial domestic consumption, and its export potential for gas is limited, Turkmenistan is mostly important for the analysis of this article.

Since the USSR disintegrated, Turkmenistan has been suffering from its landlocked position. Its attempts to build alternative gas pipelines have not been successful[17], and it has had to negotiate the price of its gas with Russia (who remained a

17 Olcott M.B. (2006) International gas trade in Central Asia: Turkmenistan, Iran, Russia and Afghanistan. In: *Natural Gas and Geopolitics: From 1970 to 2040*, D.Victor, A.Jaffe, M.Hayes, Eds.; Cambridge University Press, pp. 202-233.

monopolist for its gas transit) (for formal model of this game see Yegorov & Wirl[18]).

What is important for us in this article? The existence of the USSR before 1991 gives a unique chance to see the perspective of a central planner. Because of the relatively heterogeneous resources of natural gas in different locations of this vast country, the planners rationally[19] started extracting the cheaper resources before going on to the more expensive resources. Thus, the cheap cost (compared to the reserves in the North of Russia) of extracting natural gas from Turkmenistan is the first stylized fact. And here is another confirmation. Super giant field Gunotra Yoloten requires $10 bn of investment to produce at a capacity of 80 bcm/yr by 2024[20], and these investments are peanuts in comparison with the required investment of $200 bn in the Yamal or Shtokman super giant fields in Russia's North. But why then did investors not come up with such huge amounts of money after independence? This is where geography and politics come into play: after the dissolution of the USSR, gas transit became the core problem. As for the Russian oligarchs, it was probably another mistake not to privatize the gas resources of Turkmenistan in the 1990s, since there were plenty of more efficient projects at the time. The current policy of the Turkmen government is not to allow foreign investors access to the giant onshore fields and to try and auction off the offshore fields. Here, in the Caspian Sea, there are several problems. One is related to the unresolved dispute regarding the division of the bottom of the Caspian Sea among 5 states. Another is the necessity for advanced technology that Turkmenistan might not have. And the major problem is that Russia and Iran are formally blocking the construction of the Trans-Caspian gas pipeline on environmental grounds.

In recent years, Russia has monopolized the transit of Turkmen gas. Recently, however, China broke this monopoly by investing directly in fields and pipelines. This may be one reason why Turkmenistan has been able to substantially raise its gas price at the border, from $65/bcm in 2006 to in $ 130/bcm in 2008.

The history of gas trade between Turkmenistan and Russia spans several periods. After gaining independence in 1991, Turkmenistan tried to enforce Western prices, but lost demand and experienced payment defaults. In 1995 it

18 Yegorov Y., Wirl F. (2010b) Gas Supply Security, Competition and Geopolitics, presentation at the 11th IAEE European conference, Vilnius (Lithuania), 25-28 August 2010 (paper on-line http://www.iaee2010.org/sites/default/ufiles/yury.egorov_1.pdf).

19 Clearly, the centrally planned system is correctly criticized for its inability to deal optimally with many economic issues, especially with consumer goods. But this is an opposite case, where rationality is relatively easy.

20 Pirani S., Ed. (2009) Russian and CIS Gas Markets and Their Implications on Europe, Oxford University Press.

changed its policy, adopting a fixed price of about $40/tcm at the border and enforced payment discipline, but lost trade volume. After 2000, Turkmenistan abolished bartering, and its gas prices started to rise (more in line with world prices, but still well below them). Russia partly supplied its gas to the Ukraine, when prices in the Ukraine were well below prices in Europe.

The present gas agreement was signed between "Gazprom" and President Berdymukhammedov in 2008. There is a long term fixed-price formula with the possibility of renegotiation. Clearly, Turkmenistan's bargaining power depends on the existence of alternative transit routes, and the first alternative is coming from China. Since 2009, the price at the Turkmen border has been linked with European netback, initially at values in the range of $200-300 per tcm. However, the European gas price has dropped today, and has pushed this price down. By setting this option of buying Turkmen gas at a high price, Russia secures its position as a major transit country for Turkmen gas. Indeed, even if the Trans-Caspian pipeline is built, the European netback price offer for Turkmenistan will be lower (the figure of $175 is presented in[21]). The agreement with Russia also includes raising the capacity of the Caspian Littoral Pipeline (CLP) to 30 bcm/yr and Russia's participation in field development. Under this agreement, Russia has an option to invest substantially in fields and the expansion of transit. Given the existing pipeline network, it can keep the netback price margin at its competitors' level (most likely exceeding $50 per tcm) and exploit its pipeline system without much investment, but with a possibility of gaining substantial transit revenues by expanding its volume of transit.

What may happen in this "frozen scenario"? If no other pipelines are built, Russia would be interested in the expansion of Turkmen gas exports, since Russia also earns profits in the shape of transit fees[22]. And then it may use some gas from Turkmenistan to supply its contracts with Europe, postponing investment in Northern fields.

---

21 Pirani S., Ed. (2009) Russian and CIS Gas Markets and Their Implications on Europe, Oxford University Press.

22 Russia will face a "competitive transit fee" when rival pipeline Nabucco is built, since both will be able to deliver gas from the same source region (Caspian) to the same destination (EU). However, Nabucco will recover its investment costs and make marginal profits, while Russia will exploit the Soviet gas pipeline system without paying dividends to the non-existing USSR, and thus enjoying some rent.

5. The Model

We start from a static model, and move to dynamic. The basic idea is that in the given environment (world price for gas, bargaining share, costs of extraction) it might be more profitable to either rely on transit profits from Central Asian gas, or to invest in the development of the more costly Northern fields. Depending on the required export volume, it might be optimal to utilize both strategies. However, below a certain price threshold it might become suboptimal to subsidize costly Northern development.

The dynamic setup might bring a different picture. Under certain conditions, it may be optimal to have temporary losses, in order to preserve future market share, when gas will be more expensive. In a similar manner, investment in geopolitical power in the transit games with Central Asian countries can prolong the quasi-monopolistic position in terms of their gas transit and generate a higher level of profits from bargaining.

Let us introduce the following notations: p – gas price, Y – quantity produced in new Yamal fields, Q – quantity produced in Central Asia and the Caspian zone, E – gas demand of Europe (or long term contract obligations of Russia), B- bargaining share of Russia, C(Q), C(Y) – cost functions for Southern and Northern gas.

The model assumptions are as follows (see next section for justification): Russia produces all of the gas for export to Europe in the Northern fields and Central Asia. Production expansion in the North (replacement of Nadym-Pur-Taz (NPT) by Yamal) is costly, while production in Central Asia is much less costly. However, the bargaining share of Russia for this gas (what it can get for transit) depends on its regional geopolitical power (that has a natural tendency to decline without geopolitical investments) and on the necessity to satisfy long term contracts (the more this gas is necessary, the less bargaining share Russia is willing to accept). Suppose that

$$C(Y) = bY+Y^2,\ C(Q)=aQ,\ a<b. \qquad (1)$$

Then, for a given world (European) gas price p, the profit of Russia would be:

$$\Pi = [pY - C(Y)] + B[pQ - C(Q)]. \qquad (2)$$

If B were constant, everything would depend on the relationship between price and cost. For a very low price p, Russia would only re-export Caspian gas (since its profit is linear in volume), given there is no constraint in production. An intermediate case, when part of the gas comes from Yamal and part is Caspian, is a more real situation. Here the volume Y is determined by equality in marginal costs, while Q=E–Y.

5.1 The case of constant bargaining power

If the bargaining power is constant, Russia maximizes its profit function, taking into account the constraint on Y and Q:

$$\text{Max}_Y\, \Pi,\ \Pi=Y(p\text{-}b\text{-}Y)+B(E\text{-}Y)(p\text{-}a). \qquad (3)$$

In this case the optimal output is:

$$Y^* = \tfrac{1}{2}\,[p\text{-}\ b\text{-}\ B\ (p\text{-}a)]. \qquad (4)$$

It is easy to check the comparative statics:

$$\partial Y^*/\partial p>0,\ \partial Y^*/\partial a>0,\ \partial Y^*/\partial B>0,\ \partial Y^*/\partial b>0. \qquad (5)$$

Proposition 1. If the bargaining power of Russia with Central Asia rises or if the cost of its gas extraction rises, it will produce less domestic gas, substituting it with re-exported gas from Central Asia. If the price of gas in Europe or the price of extraction in Central Asia rises, Russia will produce more domestic gas to export to Europe.

5.2 The case of limited capacity of Central Asia

Some small extensions (reflecting real situations) can be easily done. It also may happen that Q is limited by some constraint. Production capacity in the Caspian region may be limited at certain times, as may the volume of gas left for Russian transit after a certain part has been taken by alternative importers (like China or Nabucco[23]).

Here we formally assume that there exists a constraint $Q<R$, that becomes binding, i.e., we have $E\text{-}Y^*>R$ in the previous case. Now the Russian profit function and optimal output are given by:

$$\Pi_2 = Y\ (p\text{-}b\text{-}Y) + B\ R\ (p\text{-}a),\ Y_2= \tfrac{1}{2}\ (p\text{-}b) >Y^*. \qquad (6)$$

Now Russia will produce a higher output ($Y_2>Y^*$), and its properties are as follows.

Proposition 2. If the possibility of Central Asia to transit gas via Russia is less than would be optimal for Russia, then we have a corner solution where Russian

23 If this gas is taken by Nabucco, then it also changes European demand for Russian gas.

gas production for export will no longer depend on its bargaining power, but only on its price and the cost of extraction in Russia.[24]

5.3 The case of variable bargaining power

Finally, we can consider B=B(X), X=E–Y, B'(X) <0. The following formula is relatively simple and reflects the natural decline of Russian bargaining power when it needs more gas from Central Asia to satisfy its contracts with Europe:

$$B = 1-(E-Y)/R. \quad (7)$$

In this case,

$$\Pi_2 = Y\,(p-b-Y)+(p-a)(R-E+Y), \quad (8)$$

and the optimal output is:

$$Y_2 = p - \tfrac{1}{2}\,(b+a). \quad (9)$$

This result holds when there is a resource constraint for Central Asia[25]. Now Russian output depends on the price of gas and variable costs of gas extraction in Russia and Central Asia.

These simple models show the high sensitivity of the optimal policy for gas production in Russia based on its bargaining game for Central Asian gas and the possibility to influence gas production there through investment.

6. Policy Implications. Scenario by 2020

There are many scenarios for the future, with different levels of optimism (and pessimism) for producers (and consumers). Let us start with the optimistic scenario for European gas consumers. The low price for gas will persist, and the separation from the oil price will make it independent of oil price shocks. Some cheap shale gas will be produced in Europe. The EU will progress with energy saving technologies and the use of renewable energies, and European demand for gas imports will grow modestly. For a producer (Russia) this will be bad news, since gas revenues will be low, as will investment possibilities; it also may think about economic diversification. In the pessimistic scenario for consumers, the oil peak occurs in the next years, prices sky rocket, and there is no technological

24 Corner solution is a mathematical term to denote the case when the optimal solution is on the border of a feasible region.

25 The case of no resource constraint can also be considered.

possibility of fast substitution. More gas is used as a substitute in transport, which also pushes gas prices up.

The real outcome may lie in between. Note that the huge variance in perceptions of the future adds additional risk for investors, and "freezes" some of them. Geopolitics are a further reason for "freezing" projects; we have seen a lot of such examples in the past.[26]

That is why it makes sense to consider the so called "frozen scenario", which is important for further policy implications. We know that initially economically profitable plans in gas markets are often postponed for political and geopolitical reasons. Some plans only move forward if they serve the mutual interests of potentially conflicting groups. We have seen that plans to build southern gas pipelines from Turkmenistan emerged almost 2 decades ago, but have still never been implemented. And even if the plans are implemented, they may not function as well as envisaged, nor produce the expected economic benefits. That is why it is important to consider such major plans and to describe the economic and other incentives of the different participants. Such analysis has been suggested in the paper by Hartley and Medlock (2009) where some scenarios are investigated with respect to several uncertainties, like the construction of an East-West connector (gas pipeline) in Russia.

At present, we can see several plans of this type. The first group is related to building alternative corridors for Caspian gas. The Nabucco project is competing here with South Stream. As has been suggested by Yegorov & Wirl[27], the recessionary environment after the global crisis of 2008 was not favourable for the economically profitable coexistence of both projects. Nevertheless, the construction of both can be pursued through geopolitical drive, although their profitability might be questionable over the next decade. Still, we can expect these pipelines to be built, but more for geopolitical reasons (they increase the bargaining power of both players), since they will not generate profits due to insufficient gas flows.

The situation is similar with regard to the development of new giant fields in Russia's North (Shtockman and Yamal, in particular). Given the high investment costs and no clear prospects about a favourable price for natural gas, neither Gazprom nor Russia as a state have clear incentives to proceed with fast development. At the same time, Russia needs to balance its gas production against the demand for it.

---

26 Victor, D.; Jaffe, A; Hayes, M., Eds. (2006) *Natural Gas and Geopolitics: From 1970 to 2040*, Cambridge University Press, 488 p.

27 Yegorov Y., Wirl F. (2010b) Gas Supply Security, Competition and Geopolitics, presentation at the 11th IAEE European conference, Vilnius (Lithuania), 25-28 August 2010 (paper on-line http://www.iaee2010.org/sites/default/ufiles/yury.egorov_1.pdf).

As shown by Stern[28], the Russian Gas Matrix in 2008 included the supply of 550 bcm by Gazprom, 114 bcm from other producers and 61 bcm of re-exported gas from Central Asia. At the same time, domestic consumption amounted to 353 bcm, export to Europe to 159 bcm (long term contracts only, total sales were 189 bcm) and export to CIS countries to 92 bcm. The main Russian gas field that is currently in operation is Nadym-Pur-Taz. Its natural decline between 2010 and 2020 is about 30%, or 130-150 bcm. The major replacement is being planned in the shape of development of the Yamal peninsula.[29] The development of other regions is more smooth, and thus Russia's best chance of replacing and expanding its gas production over the next decade lies in investing in Yamal. In this scenario, Gazprom would increase its production to 570 bcm by 2020. Reports also suggest the possibility of flexible production expansion, depending on incentives. But in any case we can see that expansion is subject to costly investment.

However, given the policy of moving to European netback prices (that is, European price minis tax minus transport cost) for gas sales in domestic markets and to CIS countries[30], there is also a possibility of shrinking domestic and CIS demand[31]. In this case, the increase of re-exported gas from Central Asia can be a potential answer to the positive demand shock from Europe (to keep the EU quiet about Russia's possibility to satisfy its long term contracts). Indeed, even if the Yamal development moves 50% slower (bringing not 140 bcm of additional gas by 2020, but only 70 bcm), Gazprom's production will still remain at the level of 500 bcm without substantial investment in new costly fields in the North. Given the Russian Gas Matrix, such a decline can be compensated by an increase in re-exported gas from Central Asia. Thus, we see that our model can indeed be applied realistically.

In summary, it is possible to say that – in the "frozen scenario" – Gazprom will not invest much in the North (waiting for sustainable EU gas demand for its gas at reasonable prices by 2020), but will try to compensate its declining fields by strengthening (or just maintaining) its bargaining position for Central Asian gas,

28 Stern J. (2009) Future Gas Production in Russia: is the concern about lack of investment justified? – Oxford Institute for Energy Studies, NG 35, 17 p.

29 Pirani S., Ed. (2009) Russian and CIS Gas Markets and Their Implications on Europe, Oxford University Press.

30 Pirani S., Ed. (2009) Russian and CIS Gas Markets and Their Implications on Europe, Oxford University Press.

31 A substantial decline in gas consumption by the Russian population is not likely to take place for political reasons, despite the existing positive trend in domestic prices. A very high proportion of the Russian population is less wealthy (due to high income dispersion) and demand for heating is inelastic. That is why raising prices to world levels is politically not acceptable for the Russian government, if it wants to keep the country stable and avoid a mass redistribution of income in favour of the poor.

with either direct or indirect investment in the expansion of production there. At the same time, it will use the shift to EU netback prices in its domestic market to reduce domestic demand somewhat to compensate for the natural decline.

Obviously, the growing competition for Caspian gas from China and the EU (Nabucco et al) will not allow for this strategy to become a long term policy, but it does offer some respite to escape the necessity of having to invest soon under less favourable conditions.

At the same time, what will Europe do? At present, the EU is pursuing a harsh policy towards Russia, motivated by its desire to increase energy security. However, at has been shown by[32], this short term security will result in long term insecurity as production of alternative suppliers declines. For example, researchers from Uppsala University expect Norwegian gas production to peak between 2014 and 2020. Norwegian gas plays an extremely important role for Europe: it exports 97 bcm there, which is 18% of the EU's gas supply and 30% of its gas imports.[33] Being the second largest gas exporter to the EU after Russia, Norway is Russia's biggest competitor. Domestic gas production in the EU is already in decline, while the prospects of shale gas are likely to be overestimated. As shown by Kuhn[34], the production costs of shale gas in Europe would substantially exceed those in the USA even if the technology were fully adapted: the additional costs will come from environmental damage and the different laws on land property.

All these factors will change the fundamentals for pricing gas in Europe. In the short run demand will rise, and the present spot price, which is lower than the oil-backed price for long term contracts, will grow. Insufficient investment in Russia will not allow for rapid expansion of the gas supply to Europe, and prices may reach levels close to those witnessed in 2008. If oil starts to show signs of peaking, the oil-backed gas price will reach such high levels that, on the one hand, investment in Russia will become profitable, while on the other hand, Europe will be motivated to further substitute hydrocarbons with renewable energies. After the completion of the 20-20-20 programme (by 2020) a 40-40-40-type follow-up programme is likely to emerge, which will provide completely different incentives for Russia for the period beyond 2020, and fast development, not only of Yamal but also of Shtockman, is likely to start.

---

32 Yegorov Y., Wirl F. (2008) Energy relations between Russia and EU with emphasis on natural gas, *OPEC Energy Review*, 2008, v. 32, iss. 4, p. 301-322.

33 Gatermann, Reiner (2010) The Uncertain Prospects of Norwegian Gas. – *European Energy Review*, 16 December.

34 Kuhn M. (2010) Shale Gas: A Game Changer with Implications for Europe. – 11th IAEE Conference, Vilnius, 25-28 August 2010. (http://www.iaee2010.org/sites/default/ufiles/Maximilian_Kuhn_20100819.PDF.

7. Conclusions

Our analysis is based on different analytical studies related to the technical possibilities of expanding Russian production of natural gas, its cost and required investments, as well as the studies about the geopolitical game for access to Caspian gas resources. On the basis of a simple microeconomic model, based on stylized analytical facts about the gas markets, we can demonstrate that current European policy towards Russia (driven by short term interests in respect of EU energy security) can corner Russia, if the European gas price remains relatively low and sufficient volumes are not contracted over the long term, since the development of new giant fields in Russia's North requires massive investments that will only generate positive net present value if the risks are sufficiently low.

In light of the current EU policy towards Russia (or what is named "European energy security" and includes medium-term attempts to diversify European gas imports at the expense of the Russian share and to keep gas prices low), we have shown that Russia is motivated to play the geopolitical game for Caspian gas more aggressively.

We have also investigated the "frozen scenario", where high risks postpone Russian investment in new fields and new gas pipelines from the Caspian area. In this case, Russia can save a lot on investment, by replacing its decline from the depreciated fields with larger volumes of transit gas from Central Asia, which costs much less, given the already existing pipeline network (inherited from the USSR) and allows Russia to earn substantial revenues even at competitive transit prices. In this case, Russia can substitute the very high investment cost in the North with much cheaper investment in geopolitical power in the South, thus fulfilling its contract obligations with Europe. This can happen (and under certain conditions will be optimal for Russia) even if the EU completes its alternative projects, like Nabucco. Meanwhile, the price for gas will rise, and investments in the North will become economically viable. Thus, Europe's mistakenly short-term approach to security policy will incur additional investment by both sides in the geopolitical struggle for Caspian resources. At the same time, giant fields will be developed there before any less efficient investment is channelled to Russia's North, which is just the decision that would be taken from the perspective of a central planner, as well.

Is a cooperative scenario possible? In other words, can Russia and the EU agree on a bilateral agreement and investment in Caspian gas (and potentially later in Russia's North), so that losses are minimized? It would be a favourable economic outcome. This can be proven using the method of backward induction to reach an equilibrium in game theory. And this article attempts to hint in this direction. However, politicians rarely depart from their ambitions (and non-cooperative games are not played "cooperatively").

8. Literature

BP Statistical Review (2008)

Ericson R. (2009) Eurasian Natural Gas Pipelines: The Political Economy of Network Interdependence. – *Eurasian Geography and Economics*, v. 50, No. 1, p. 28-57.

Finon D. (2009) The limits of the EU direct foreign gas policy. Autopsy of the stillborn Southern corridor project Nabucco, CIRED WP, December 2009, 17p.

Holz F., von Hirschhausen C., Kemfert C. (2009) Perspectives of the European Natural Gas Markets Until 2025. – *The Energy Journal*, Special Issue. World Natural Gas Market and Trade: A Multi-modelling perspective, p. 137-150.

Hartley P., Medlock K. (2009) Potential Futures for Russian Natural Gas Export. – *The Energy Journal*, Special Issue. World Natural Gas Market and Trade: A Multi-modelling perspective, p. 73-95.

Hubert F., Ikonnikova S. (2004) Hold-Up, Multilateral Bargaining, and Strategic Investment: The Eurasian Supply Chain for Natural Gas. – WP, Humboldt University, Berlin.

Gatermann, Reiner (2010) The Uncertain Prospects of Norwegian Gas. – *European Energy Review*, 16 December.

Kuhn M. (2010) Shale Gas: A Game Changer with Implications for Europe. – 11th IAEE Conference, Vilnius, 25-28 August 2010. (http://www.iaee2010.org/sites/default/ufiles/Maximilian_Kuhn_20100819.PDF.

Olcott M.B. (2006) International gas trade in Central Asia: Turkmenistan, Iran, Russia and Afghanistan. In: *Natural Gas and Geopolitics: From 1970 to 2040*, D.Victor, A.Jaffe, M.Hayes, Eds.; Cambridge University Press, pp. 202-233.

Olcott M. (2009) Russia, Central Asia and the Caspian: How Important is the Energy and Security Trade-off? – WP, James Baker III Institute for Public Policy of Rice University, USA, 31 p.

Pirani S., Ed. (2009) Russian and CIS Gas Markets and Their Implications on Europe, Oxford University Press.

Stern J. (2009) Future Gas Production in Russia: is the concern about lack of investment justified? – Oxford Institute for Energy Studies, NG 35, 17 p.

Ten Hoedt, Rudolf (2010) Interview: Azerbaijan's Top Negotiator Elshad Nassirov. – *European Energy Review*, 15 November.

Victor, D.; Jaffe, A; Hayes, M., Eds. (2006) *Natural Gas and Geopolitics: From 1970 to 2040*, Cambridge University Press, 488 p.

Yegorov Y., Wirl F. (2008) Energy relations between Russia and EU with emphasis on natural gas, *OPEC Energy Review*, 2008, v. 32, iss. 4, p. 301-322.
Yegorov Y., Wirl F. (2010a) Gas Transit, Geopolitics and Emergence of Games with Application to CIS Countries, USAEE – IAEE WP No. 10-044, February 2010, 29 p. (http://ssrn.com/abstract=1560563).
Yegorov Y., Wirl F. (2010b) Gas Supply Security, Competition and Geopolitics, presentation at the 11th IAEE European conference, Vilnius (Lithuania), 25-28 August 2010 (paper on-line http://www.iaee2010.org/sites/default/ufiles/yury.egorov_1.pdf).

Regional Integration and Energy: The Eurasian Case

Yeşim Reel

1. Introduction

The processes of regional integration that emerged after World War II were originally mostly related to trade and economics. Especially since the 1980s, however, with the so-called wave of 'new regionalism', regional integration has become what is regarded as a multidimensional process. Indeed, the number of regional integration initiatives is steadily increasing.[1] Some of them are seen in Eurasia.

It is generally accepted that Eurasia was as an integrated economic space during the first half of the last millennium, created by Ghengis Khan and his heirs starting in the 12th century and symbolized by the Great Silk Road. The political disintegration of the Mongol Empire, however, coupled with the growing ease of sea transport and the rise of colonial competition led to the disintegration of this integrated Eurasian space, starting about the time that Christopher Columbus and his fellow explorers began a centuries-long process of transatlantic and later transpacific integration. More recently, the self-imposed economic isolation of the Soviet Union and Communist China meant that large parts of Eurasia did not participate in the "globalization" of the world economy after the Second World War.

As China opened up to the rest of the world in the 1980s and the Soviet Union collapsed, political barriers fell away, and strong forces of economic integration have now taken hold in Eurasia. Integration is seen to be proceeding very rapidly across a wide spectrum of economic and human interaction. The most obvious area being that of energy.[2] Energy trade is considered the single most important and rapidly growing economic link for Eurasia.[3] In fact, it is suggested that the integration of the Eurasian super-continent has potentially major impli-

1 De Lombaerde, P. and Langenhove, L. V. (2005). Indicator of Regional Integration: Conceptual and Methodological Issues, *IIIS Discussion Paper* No. 64, p. 3.

2 Linn, J.F. (2006a). Eurasian Integration – a Super-continental Opportunity for Russia, Wolfensohn Center for Development, Kommersant November 24. At http://www.brookings.edu/opinions/2006/ 1124russia_ linn. aspx?p=1 (18/02/2011).

3 Linn, J.F. (2006b). *The New Impetus towards Economic Integration between Europe and Asia*, at The Fourteenth OSCE Economic Forum, Vienna 23-24 January 2006, EF.NGO/1/06, p. 4.

cations for the Eurasian region, and also for the world economy.[4] For these reasons, it is important to work on Eurasian integration and especially on energy trade.

Investigating issues around this, the present chapter is composed of four sections. Following this introduction, the next section presents an overview of regional integration, with a brief definition. The third section examines the Eurasian case, with general information about Eurasian integration and its importance. This consists of two sub-sections, the first looking at the energy trade in Eurasia, and the second detailing the current situation and energy potentials in Eurasia. The final section gives some concluding remarks.

2. Regional Integration

Regional integration is stated to be a worldwide phenomenon of territorial systems that increases the interactions between their components and creates new forms of organization. The processes of regional integration that emerged after World War II, and developed with the so-called 'new regionalism' of the 1980s into what can be seen as a multidimensional process imply – besides economic cooperation – also dimensions of politics, diplomacy, security, culture, etc. Nevertheless, trade and economic integration remain a central aspect of ongoing integration schemes. Indeed, the number of regional integration initiatives is steadily increasing.[5] Some of them are in Eurasia.

Regional integration is defined along three dimensions:

(i) **Geographic scope** – concerning the number of countries involved in an arrangement,

(ii) **Substantive coverage** or width – related to the sector or activity coverage (trade, labour mobility, macro-policies, sector policies, etc.),

(iii) **Depth of integration** – measuring the degree of sovereignty a country is ready to surrender, that is, from simple coordination or cooperation to deep integration.[6]

---

4 Linn, J.F. and Tiomkin, D. (2007). "Economic Integration of Eurasia – Opportunities and Challenges of Global Significance", in Anders Aslund and Marek Dabrowski, (eds.) Europe After Enlargement: Newyork, Cambridge University Pres, p. 162.

5 De Lombaerde, P., and Langenhove, L.V., p. 3.

6 Kritzinger, L. (2005). Regional Integration: Concepts, Advantages, Disadvantages and Lessons of Experience, May 2005, p. 4. At www.sarpn.org.za/documents/.../P1416-RI-concepts_May 2005.pdf.

Regionalization is one way to respond efficiently to the challenge of globalization. Countries in a certain region share relative national identities and interests. The process of integration through which these regional agendas and identities are formed and sustained to facilitate cooperation is called regionalization. It is enhanced through the integration of several dimensions including economic, social and political relations. Economic integration can be classified in four stages according to the degree of cooperation. The free trade agreement (FTA) is the first, in which tariffs are reduced to zero among related parties. The next stage is customs unions, in which no tariffs are imposed among member states and common external tariffs are levied against outside countries. The third stage is a single market in which the free movement of goods, services, capital, and people are guaranteed, and finally the fourth stage is an economic union in which legally binding, common economic policy is realized among member states. Economic integration is based on mutual interest.[7] Some of these integration stages are seen as different instances in Eurasia.

3. The Eurasian Case

The ending of China's isolationist policy following Kissinger's diplomacy and the collapse of the Soviet Bloc after the fall of the Berlin Wall led to the opening up of previously closed borders and signalled the transition from central planning to market economies, not only in ex-socialist countries, but also in the West and the world generally, where the 1989 Washington Consensus marked the arrival of the contemporary period of neo-liberalism. This heralded a historic new phase of Eurasian development by permitting and encouraging the integration of national economies with each other, including those of Eurasia. This process of integration is now connecting some of the largest and most dynamic economies of the world – China, India, Russia – with each other and with an expanding European Union.[8]

As Ferguson (2006) has stated, Eurasia – geographically comprising the interaction of Europe, Russia, Central Asia and the Far East – has gone through dynamic political and economic changes that affect global politics and adjacent regions. If not yet an integrated super-region, it has been suggested that Eurasia is a new 'convergence zone', where the interests of a number of major and medium-size powers are interacting. Eurasian convergence in the widest sense is the inter-

7 Kim, N.K. (2011). Globalization and Regional Integration in Europe and Asia, Ashgate publishing, pp. 1-2.

8 Linn, J.F. (2006b), p. 3.

action of this entire zone, ranging from Europe to the Far East, and including the regions of **Central Asia** (including Kazakhstan, Uzbekistan, Tajikistan, Kyrgyzstan, and Afghanistan)[9], and the **Caucasus**, which links the Caspian and Black Sea areas, including Armenia, Georgia, and Azerbaijan, as well as influencing parts of Iran, Turkey and South Russia. Also, a new term has begun to emerge in literature, that of **Greater Central Asia**, suggesting the Central Asian states have a strong interaction with adjacent areas in southern Siberia, Mongolia, Western China and Tibet, and Pakistan and Kashmir.[10]

Other than Afghanistan, the countries of Central Asia and the Caucasus that comprise **Central Asia** are independent states created with the break-up of the Soviet Union. Twelve of these fifteen states formed the **Commonwealth of Independent States** (CIS) during the early 90s.[11] Though still partly dependent on Russian trade and transit routes, the ex-Soviet states are keen also to develop new trading partners, including China, South Korea and Japan, as well as the EU and the US. This is particularly noted in the area of investment and technical assistance. They have also begun some level of involvement in other new **regional organizations** such as the **Black Sea Economic Cooperation** (BSEC) process (part of a process from June 1992, but upgraded to a formal organization in 1998) and the **Economic Cooperation Organization** (ECO).

The **BSEC** group has eleven members – including Greece, Turkey and Albania, six ex-Soviet countries (Russia, the Ukraine, Georgia, Armenia, Azerbaijan and Moldavia) and two from the Soviet bloc (Romania and Bulgaria) – and is 'seeking to create a market of 350 million people and to develop mutually advantageous cooperation' including the long term goal of 'greater democracy, peace and development'.[12] Established in 1985 by founder members Iran, Pakistan and Turkey, the ECO aims to create a single market.

The other important integration example is the **Eurasian Economic Community** (EurAsEC) composed of CIS member states Belarus, Kazakhstan, Kyrgyzstan, Russia, Tajikistan, and Uzbekistan. The Agreement to create EurAsEC was signed on October 10th 2000 by the leaders of five of these countries, the exception being

---

9 Afghanistan is sometimes regarded as part of South Asia: in fact is sits astride linkage points to several areas.

10 Ferguson, R.J. (2006). Lecture 1: Eurasia: Super-region or Zone of Conflict?, FHSS, Bond University, QLD Australia, INTR13/71-304 Eurasia, p. 1. At http://www.international_relations.com/wbeurasia/WBEA-2006-Lec1.doc. (22/02/2011).

11 Two of which, Georgia and the Ukraine, later withdrew, aligning themselves in a new grouping formed at the end of the decade, the Organization for Democracy and Economic Development, GUAM, an acronym of its member states, which include also Azerbaijan and Moldavia). The main practical work of this organization has been to develop plans for a transportation corridor.

12 Ferguson, R.J., p. 2.

Uzbekistan, which joined EurAsEC in 2005.[13] Moldavia, the Ukraine and Armenia are Observers of EurAsEC. EurAsEC is an international economic organization that aims to form common external customs borders and develop a common external economic policy, tariffs, prices, etc., and other functioning parts of a common market, as adapted from the European Community, WTO, and other international norms. One EurAsEC area of interest is energy, which involves enhancing common hydro-energetic complexes located in Central Asia and resolving water and electricity supply problems, in working towards a common energetic balance.[14]

The EurAsEC zone operates as a transit region for overland continental transport – East-West from China to Europe and North-South from Russia to India. Indeed, aggregate intra-Eurasian trade has already emerged as more important for the regional economies than trade with the rest of the world.[15]

3.1 Energy Trade in Eurasia

Cheap and abundant energy suppliers were like a bonding agent that kept the Soviet Union together; energy suppliers and interdependencies have also been central to the processes of both separation from the centre and reintegration in the former USSR (Union of Soviet Socialist Republics).

For many of the former Soviet Republics, energy is literally the number one issue – it directly affects their daily economic situation, their domestic politics, and the relationship with their main international partners. This is especially so for those eight energy-poor states which remain largely dependent on imports from Russia: Belarus, the three Baltic States, Moldavia, Armenia, Georgia, and last but not least the Ukraine. Thus, energy dependency emerges as a central reality in the lives of most post-Soviet states.[16]

Energy issues have been central both for conflict in the former USSR and for Russian-led reintegration attempts. In the post-Soviet region, energy issues have been closely related to conflict at both inter-state and domestic levels: conflicts over access to oil and gas markets and supplies; conflicts over energy transit; and conflicts over the distribution of economic gains – and, conversely, the appor-

13 Uzbekistan suspended its membership in 2008.

14 Eurasian Economic Center. At http://www.eurasian-ec.com/index.php?option=com_content&task=view&id=2&Itemid=7. (15/03/2011).

15 Norling, N. and Swanström, N. (2007). "The Virtues and Potential Gains of Continental Trade in Eurasia", *Asian Survey*, 47 (3): pp. 351-352.

16 Balmaceda, M.M. (2007). Energy, Integration and Disintegration in the former USSR, Routledge, p. 1.

tionment of losses – from energy trade, transit, and domestic sales. In many of the post-Soviet states, it is not a case of national states domestically united in fighting to overcome their energy dependency on Russia, but rather a multiplicity of competing interest groups, both domestic and foreign, all trying to make a profit out of this situation of dependency. In fact, it is suggested that this is true of all the issues related to energy in the process of post-Soviet dissolution and reintegration.[17]

Energy trade is considered the single most important and rapidly growing economic link for Eurasia. The region (excluding the Arab Peninsula) has 25% of the world's proven oil reserves, 36% of oil production, and 55% of oil consumption. Regarding gas, the region has 57% of world reserves, 50% of production, and 57% of consumption. World energy trade, while integrated across oceans into a single market, is actually quite heavily concentrated in the primary continental masses – the Americas and Eurasia –, with only limited cross-oceanic trade in oil and virtually none in gas and power.

A highly integrated network of oil and gas pipelines now links Western and Central European consumers with Russian, Iranian and Caspian oil and gas fields, and is continuing to develop. More recently, pipelines have also been initiated or begun to be considered from these key supply sources to China and Japan, as well as to the Indian subcontinent. Ambitious plans are also under consideration for the development of the hydro resources of Central Asia and the transmission of electricity to consumers in Afghanistan, China, India, Russia and even Central and Western Europe.[18]

Given the dynamic demand and supply outlook, regional energy trade is considered a major force for integration in Eurasia. But there are also major challenges to be met in order for this trade and integration process to unfold smoothly. First, major investments in production and transportation have to be made to overcome the geographical constraints facing energy production and trade in the region. Second, financing constraints can also slow down expansion, with innovative public-private partnerships required to assure an adequate source of funding for very large investments. Third, security concerns in parts of the region, especially Afghanistan and the long-contested Indian-Pakistani border, may interfere with energy transmission. Fourth, a lack of integrated regulatory structures for the whole region can present important obstacles. And finally, political competition for dedicated supplies (e.g. between China and Japan), or Russian efforts to maintain a monopoly over energy transport routes and to use

17 Balmaceda, M.M., p. 2.

18 Linn, J.F., pp. 4-5.

them for political purposes, may get in the way of economically rational expansion of energy trade in the region.[19]

The rewards of overcoming these challenges promise to be considerable, however. Opening Greater Central Asia to continental trade in energy and goods would allow countries in the region extensive access to foreign technology and foreign exchange revenue, and increase market access.[20] There are also enormous potential yields in energy transit. In particular, if the energy demands of India and Pakistan are met by building the proposed Turkmenistan-Afghanistan-Pakistan-India pipeline across Afghanistan, or the Iran-India-Pakistan pipeline, the transit states will benefit significantly. For example, it has been projected that Pakistan would gain a total of $14 billion in 30 years from the Indo-Iran pipeline.[21]

3.2 The Current Situation and Potentials of Energy in Eurasia

It can be suggested that five Eurasian countries –Russia, Azerbaijan, Kazakhstan, the Ukraine, and Turkmenistan, which are the largest consumers and producers of energy – have a distinguished role in the region. If we examine Russia first, this country is home to the world's largest natural gas reserves, the second largest coal reserves, and the eighth largest oil reserves. It is also the world's largest exporter of natural gas and second largest oil exporter (and the third largest energy consumer). Although Russia has tremendous natural gas reserves, the industry will encounter serious investment challenges in opening new fields. Russia is a major oil producer, sometimes producing even more than Saudi Arabia.[22]

Russian energy policy is a key issue in global energy policy, and energy policy is crucial for Russia. The subject of Russian energy is as complex as it is crucial. This complexity has made Russian gas and oil, in particular, together with the proceeds from the export of these goods and the associated politics, a frequent and divisive topic in academic analysis and policy commentary.

The oil and gas fuelled agenda of Russian-European and Russian-Eurasia energy relations. The most familiar being how Russia has become an increasingly important energy supplier to its fellow BRIC (Brazil, Russia, India and China) member China, as well as Japan and South Korea. Of these new customers, China became the world's biggest consumer in 2009. This new market conquest is making Russia a player in Asia with a weight not seen since the dissolution of the Soviet

19 Linn, J.F., pp. 6-7.
20 Norling, N. and Swanström, N., p. 369.
21 Pandian, S. G. (2005). "Energy Trade as a Confidence Building Measure between India and Pakistan: A Study of the Indo-Iran Pipeline Project", *Contemporary South Asia*, 14 (3): p. 314.
22 IEA (2010a). International Energy Annual, **At** http://www.eia.gov/countries/cab.cfm?fips=RS. (05.08.2010).

Union in 1991. Russian energy companies have also established partnerships with other energy producers in Asia, Africa and America.[23]

Three of the eight post-Soviet states in Central Asia and the South Caucasus regions are significant exporters of oil or natural gas: Azerbaijan, Kazakhstan and Turkmenistan. Since 2003, high petroleum income has provided these three countries with an opportunity to stimulate economic development, improve the living standards of their populations and expand their roles on the international stage. However, as the experience of many other resource-rich countries has shown, having abundant oil and natural gas resources also presents serious economic and political challenges. Several unique characteristics of petroleum wealth mean that governments that are heavily reliant on this source of income must undertake certain tasks that other governments largely ignore. This includes pursuing careful economic policies to offset the volatility of international oil prices and to compensate for the eventual depletion of their petroleum reserves.[24]

The collapse of oil prices due to the global financial crisis in 2008-2009 created further challenges for oil and gas exporters. Not only did these governments have to deal with sudden windfalls from the price hike between 2003 and 2008, they subsequently had to cope with a sudden drop in revenue.

While the economic challenges associated with petroleum abundance are significant, policymakers can take steps to alleviate them. For instance, governments can invest petroleum revenue in infrastructure and sustainable assets aimed at promoting the non-petroleum sectors of the economy and/or save these revenues abroad in order to avoid inflation and currency appreciation.[25]

Since the collapse of the Soviet Union, most of the former Soviet Republics have been striving to become part of the global economy. Among these states, Azerbaijan's performance in terms of attracting foreign investment has been the most impressive. The inward Foreign Direct Investment (FDI) Performance Index, which comprises approximately 140 countries, ranked Azerbaijan third highest from 1994-1996 and eighth from 1998-2000. Since the "contract of the century" in September 1994, Azerbaijan has signed more than twenty oil contracts with 33 companies from 15 countries. The untapped oil and gas reserves of Azerbaijan – onshore and offshore in the Caspian Sea region – and Azerbaijan's strategic location certainly explain the initial interest of international oil companies. Azerbaijan has approximately 7 billion barrels of proven oil reser-

---

23 Aalto, P. (2012). Russia's Energy Policies: National, Interregional and Global Levels, Edward Elgar, pp. 3-4.

24 Overland, I. K., Herschman, H., Kendall-Taylor, A. (2009). "The Resource Course and Authoritarianism in the Caspian Petro States", in Indra Kjaernet Overland, Heidi Herschman, Andrea Kendall-Taylor, (eds.) Caspian Energy Polics: Azerbaijan, Kazakhstan and Turkmenistan, Routledge, p. 1.

25 Overland, I.K., Herschman, H., Kendall-Taylor, A., p. 2.

ves.[26] With the start-up of the Shah Deniz natural gas and condensate field in 2007, Azerbaijan went from being a net natural gas importer to a net exporter. Indeed, Azerbaijan is emerging as a key player in the European attempt to diversify supply with alternatives to the supply routes originating from Russia. The Nabucco pipeline through Turkey, due to open in 2017, will start at Baku (and ultimately, perhaps, in Turkmenistan) as well as Iraq (the main supplier).[27]

During the last 15 years the experience of economic development in the three countries Azerbaijan, Kazakhstan and Turkmenistan has been both similar and, at the same time, different. The similarity lies in strong natural resource dependence (in FDI and export terms) of all three economies as well as their attempt to manage the resource wealth through resource funds, the establishment of which took place recently. In addition, hydrocarbon wealth became a saviour of the economies of these countries after the collapse of the Soviet economy and the transition shock. As well as similarities, many fundamental differences have been identified in the process of the analysis. These include levels of income and development, with Kazakhstan being richer and more advanced in transition than Azerbaijan and Turkmenistan.[28]

Full development of its major oilfields could make Kazakhstan one of the world's top 5 oil producers within the next decade. With production of 1.54 million barrels per day (bbl/d) in 2009, Kazakhstan is already a major producer. It is expected to at least double its current production by 2019. With large amounts of associated natural gas at its oil fields, Kazakhstan has the potential to become a net exporter in coming years.[29]

Kazakhstan is interested in the construction of a new oil pipeline to Europe, which will go through Caucasus territory and link to the Romanian Constanta-Trieste pipeline. In this case, Romania plays the role of Kazakhstan's take-off area for Europe. Kazakhstan supplies Europe with 4 million tons of oil a year, which is one third of Romania's total exports. Moreover, Kazakhstan's KazMunayGas is the 100% owner of Romanian Rompetrol, which has its own refinery and a sales network of 900 gas stations in six European countries. At present, Kazakhstan produces 70-80 million tons of oil per year, the bulk of which is being transported to Russia. Around 10 million tons go to China and 10 more million tons are transported to Baku via the Baku-Tbilisi-Ceyhan oil pipeline. In 2013, Kazakhstan intends to commence large-scale oil production on one of the world's largest oilfields, Ka-

26 Bayulgen, O. (2010). Foreign Investments and Political Regims: The Oil Sector in Azerbaijan, Russia, and Norway, Cambridge University Press, p. 87.

27 This would make Georgia, Turkey, Bulgaria, Hungary and Austria, along with Greece and Italy additional intermediary countries in the CIS-EU trade. IEA (2010b). Country Analysis Briefs – Azerbaijan, Russia, Kazakhstan. At http://www.eia.doe.gov. (27/12/2011).

28 Kalyuzhnova, Y. (2008). Economics of the Caspian Oil and Gas Wealth: Companies, Governments, Policies, Palgrave Macmillan, pp. 220-221.

29 IEA (2010b).

shagan. This would give Kazakhstan an annual "export supplement" of 50 million tons. Kashagan's reserves are expected to exceed the capacities of Russian Transneft, Russian- Kazakh CPC, Baku-Tbilisi-Ceyhan and a pipeline to China.[30]

The Ukraine is important for world energy markets because it is a substantial energy producer with valuable reserves of coal and gas, and it is a critical transit centre for exports of Russian oil and natural gas to Europe, as well as a significant energy consumer. Energy efficiency efforts and investments have been hampered by distortions in energy pricing and the energy market structure. Available estimates of future investment required to enhance energy efficiency, increase energy production and upgrade energy infrastructure vary depending on the source and the underlying energy policy priorities. As the state budget will be unable to cover more than 15% of the future energy investments, there is an urgent need to find other sources of financing.[31]

Turkmenistan has some of the largest natural gas reserves in the world, yet the country faces a myriad of challenges in bringing those reserves to world markets. It is geographically far from end-use markets and lacks sufficient pipeline infrastructure to export more hydrocarbons. The country is eager to diversify export routes for its oil and gas resources outside of the pipelines going to Russia, but must obtain capital, technical assistance, and political support for alternative pipelines.[32]

The energy data in Tables 1 to 3 provide more detailed information about Azerbaijan, Kazakhstan, Russia and Turkmenistan in this area.[33]

Table 1: Petroleum Data for Azerbaijan, Russia, Kazakhstan, Turkmenistan, Eurasia and the World in 2009

Petroleum (1000 barrels/ day)	**Azerbaijan**	**Russia**	**Kazakhstan**	**Turkmenistan**	**Eurasia**	**World**
Total Oil Production	1,012.25	9,933.71	1,540.41	198.15	12,935	84,244
Crude Oil Production	1,005.85	9,495.36	1,455.15	176.26	12,288	72,182
Consumption	130.15	2,927.00	241.00	93.00	4,320	84,684
Net Export/Imports	882.10	7,006.71	1,299.41	105.15	8,615	–
Total Oil Exports to US	75	563	29	2	732	11,691
Refinery Capacity	399	5,428	345	237	8,205	85,900
Proven Reserves (billion barrels)	7.00	60.00	30.00	0.60	99	1,342

30 Maqsimchook, N. (2011). "Chronicle of Eurasian Regional Integration 2010", in Evgeny Vinokurov (ed.) EDB Integration Yearbook 2011, Almaty: Eurasian Development Bank, pp. 277-278.

31 OECD (2011). OECD Investment Policy Reviews: Ukraine 2011, OECD Publishing, p. 98.

32 IEA (2012). Country Analysis Briefs. Turkmenistan. http://www.eia.doe.gov (28/03/2012).

33 IEA (2010c). International Energy Statistics. http://www.eia.gov/countries/cab.cfm?fips=RS (28.03.2012).

Table 2: Natural Gas Data for Azerbaijan, Russia, Kazakhstan, Turkmenistan, Eurasia and World in 2009

Natural Gas (Billion Cubic Feet)	**Azerbaijan**	**Russia**	**Kazakhstan**	**Turkmenistan**	**Eurasia**	**World**
Production	577	20,610	388	1,347	25,814	106,506
Consumption	367	15,524	304	708	21,125	107,245
Net Export/Imports(-)	209	5,085	133	636	4,734	–
Proved Reserves (Trillion Cubic Feet)	30	1,680	85	94	1,994	6,289

Table 3: Total Primary Data for Azerbaijan, Russia, Kazakhstan, Turkmenistan, Eurasia and the World in 2007

Total primary energy (10^{15} Btu)	**Azerbaijan**	**Russia**	**Kazakhstan**	**Turkmenistan**	**Eurasia**	**World**
Production	2.231	53.679	5.275	2.922	71	473
Consumption	0.591	29.969	2.320	0.924	45	482
Energy intensity (Btu/2005 USD)	9,176	15,117	14,603	60,789	15,797	7,528

The world economy demonstrated signs of recovery in 2010. The Eurasian oil and gas and energy sectors also remained a focus of attention. In 2010, oil and energy products made up the bulk of exports from Azerbaijan, Kazakhstan, Russia, Turkmenistan and Uzbekistan. Azerbaijan, which exports oil and gas, achieved a positive trade balance of $14.7 billion in 2010. As a major exporter of energy products, Kazakhstan also has a trade surplus ($29.5 billion) due to increasing export revenues. Commissioning a gas pipeline to China at the end of 2009 and a rise in raw material prices (energy products, cotton, grain) led to a slight increase in Turkmenistan's exports, which increased to $9.7 billion by year end from $9.6 billion in 2009. Uzbekistan, which joined the Turkmenistan-China gas pipeline, also benefited from this development, and thus experienced an 11% increase in exports.[34]

Encouraging energy projects in the region and regional energy integration assures long-term returns; the BTC (Baku-Tbilisi-Ceyhan) pipeline is a case in point. This pipeline not only gave Azerbaijan independence from Russia, but its construction also indicated a long-term promise from the West to Azerbaijan's prospects.[35]

34 Kurmanalieva, E., Ormantayeva, G. (2011). "Key Macroeconomic Indicators of the CIS Countries in 2010", in Evgeny Vinokurov (ed.) EDB Integration Yearbook 2011, Almaty: Eurasian Development Bank, pp. 314-315.

35 And will be further reinforced in this case with the use of the BTC line in conjunction with the Nabucco line.

Similar assistance should be provided to projects of equal weight such as the proposed Turkmenistan-Afghanistan-Pakistan-India pipeline, which could act as a symbol and milestone in Greater Central Asia relations analogous to the BTC's provision of a "window to the West."[36]

Eurasian countries desperately need investments to realize the potential offered by their vast resources. Disclosure is critical for attracting domestic and foreign investors. Insufficient disclosure is perceived as a significant risk by outside investors and can lead to higher capital costs for a company and/or reduced investment. In contrast, the enforcement of effective disclosure will make corporate insiders more exposed and responsive to external pressure and thereby enhance investor confidence, which is essential for the economic development of the region.[37] One can also argue that increased transparency in the supply chain will also facilitate the implementation of substantial investment in the sector. Eurasian countries should strive to support and guarantee the flow of supply. In the short term, there are growing concerns for investments in existing and new oilfields, and transport infrastructure for energy – especially oil and gas – and also diversifying supply and export routes.

Most investments in the energy sector and infrastructure of EurAsEC countries have been made by Russian companies. One of the most active Russian players in EurAsEC is LUKoil, a private concern formed from the merger of three state companies in 1991. LUKoil has participated in eleven oil projects in Kazakhstan since 1995.[38]

Heifetz has summarized direct Russian investments in this area, as shown in Table 5.

Russia has been significantly active in Eurasia, but its active role is expected to fade over time as new gas sources are accessed, and new technologies for both providing and using energy are developed.

On the other hand, energy efficiency is of key importance to the sustainable development of the Eurasian region, which is characterized by relatively high energy intensity. In order to implement the corresponding programmes and coordinate their efforts, development institutions are establishing strategic partnerships. Investments in energy efficiency are usually spent on the replacement or modernization of outdated production and heating facilities, the installation of metering equipment, thermal insulation, the renovation of heat and electricity

36 Norling , N. and Swanström, N., p. 370.

37 OECD (2004). Corporate Governance in Eurasia: A Comparative Overview, OECD Publishing, p. 41.

38 Heifetz, B. (2009). "Russian Direct Investments in EurAsEC and Their Role in Energy and Transport Infrastructure Development in Eurasia", in Evgeny Vinokurov (ed.), EDB Eurasian Integration Yearbook 2009, Almaty: Eurasian Development Bank, p. 130.

distribution systems, as well as different projects in the fields of biomass, biogas and solar energy.[39]

Table 5: Direct Russian Investments in the Energy Sector and Infrastructure in EurAsEC Countries.

Investor company or affiliate	**Asset**	**First transaction year**	**"Price of entry ticket" ($ million)**
Gazprom	Shakhpakhty (Uzbekistan)	2004	…
	50% of shares in KazRosGaz (Kazakhstan)	2002	…
	50% of shares in Beltransgaz (Belarus)	2007	2500
Gazpromneft	Over 100 petrol stations in Kyrgyzstan	2006	…
	Gazpromneft Tajikistan	2007	…
	Gazpromneft Kazakhstan	2007	…
LUKoil	15% in the Karachaganak PSA (KZ)	1997	…
	50% in Kumkol (Kazakhstan)	1995	…
	2.7% in Tengiz (Kazakhstan)	1997	129.1
	50% in Karakuduk (Kazakhstan)	2005-2006	
	25% in North Buzachi (Kazakhstan)	2005	…
	25% in JV Arman (Kazakhstan)	2005	
	25% in Kazakhoil-Aktobe(KZ)	2005	
	50% in the Tyub-Karagan PSA (KZ)	2004	
	50% in Atashsky (Kazakhstan)	2004	2.130
	12.5% in the South Zhambai and South Zaburunye PSAs (Kazakhstan)	2007	
	50% in JV Turgai Petroleum (Kazakhstan)	2005	
	50% of shares in Nelson Resources (KZ)	2005	
	90% in the Kandym-Khauzak-Shady PSA (Uzbekistan)	2004	…
	20% in the Aral PSA	2006	…
	100% in SNG Holding (Uzbekistan)	2008	580
Rosneft	50% in the Adaisky area (Uzbekistan)	2001	…
	RN Kazakhstan (Kazakhstan)	2003-2005	…
INTER RAO UES	50% in Ekibasruz GRES-2(Kazakhstan)	2003	…
	75% of shares in Sangtudinskaya hydro-power plant 1 (Tajikistan)	2003	…
Rusal	50% in a JV founded with Samruk (KZ)	2007	
Atomredmetzoloto	50% in Karatau LLP and 25% in JV Akbastau (Kazakhstan)	2009	560-580
Renova	Kara-Baltinsky Ore Mining Works(KG)	2007	…

Source: Heifetz, 2009, p. 141

39 Baybikova, E. (2011). "International and Regional Development Banks in Northern and Central Eurasia: Overview of Activities in 2010", in Evgeny Vinokurov (ed.) EDB Integration Yearbook 2011, Almaty: Eurasian Development Bank, pp. 323-324.

Energy demand projections assume that Eurasian electricity demand between 1990 and 2050 will increase by a factor of five in the Asian part and nearly by a factor of three in the European. The rich energy resources of Eurasia therefore need to be made available to satisfy growing demand. Resources and demand need to be matched geographically through trade and transportation networks. They should be financed through investment flows and reforms designed to attract those investment flows.[40]

The Caspian region has the potential to make a significant contribution to ensuring energy security in the rest of the world, and to increasing the diversity of global oil and gas supplies. The Caspian region contains substantial resources of both oil and natural gas, which could underpin a sizeable increase in production and exports over the next two decades. But potential barriers to the development of these resources, notably the complexities of financing and building pipelines through several countries, are expected to constrain this expansion to some degree. In the New Policies Scenario, Caspian oil production grows strongly – especially over the first 15 years of the projection period. It jumps from 2.9 mb/d in 2009 to a peak of around 5.4 mb/d between 2025 and 2030, before falling back to 5.2 mb/d by 2035. Kazakhstan contributes all of this increase. As with oil, gas exports are projected to grow rapidly, rising from less than 30 bcm in 2009 to nearly 100 bcm in 2020 and 130 bcm in 2035.

While Russia will remain a purchaser of Caspian gas, there will be greater diversity in Caspian gas trade as the region expands its access to new markets in Europe and China, which is emerging as a major new customer. Further development of a Southern corridor from Azerbaijan to Turkey and other European markets will enable larger volumes of Azerbaijan gas to move westwards.[41]

4. Conclusions

It is argued that the number of regional integration initiatives has been steadily increasing since World War II, with integration proceeding across a wide spectrum of economic and human interactions, and that some instances of this integration have been occurring in Eurasia, particularly in the area of energy. In fact,

40 Klaassen, G., Grübler, A. and Schrattenholzer, L. (1999). Towards New Energy Infrastructures in Eurasia: a background paper, Interim Report IR-99-17, December, pp. 6, 8.

41 IEA (2010d). "What does the global energy outlook to 2035 look like?", World Energy Outlook 2010 Factsheet, p. 3. At http://www.worldenergyoutlook.org/docs/weo2010/factsheets.pdf. (28/03/2012).

energy trade is considered the single most important and rapidly growing economic link for Eurasia.

Regional energy trade is seen as a reason for transnational linkage and also as a major force for integration in Eurasia. Opening Greater Central Asia to continental trade in energy and goods would allow countries in the region extensive access to foreign technology and foreign exchange revenue, increase market access, and generally stimulate economic and other activity. There are also enormous potential yields in energy transit.

In the meantime, there are major challenges. The rewards of overcoming these challenges promise, however, to be considerable.

Energy issues have been central both for conflict in the former USSR, and for Russian-led reintegration attempts. In the post-Soviet region, energy issues have been closely related to conflict at both inter-state and domestic levels. In fact, it is suggested that this is true of all the issues related to energy in the process of post-Soviet dissolution and reintegration.

Russia has been significantly active in Eurasia, but its active role is expected to fade over time as new gas sources are accessed, and new technologies for both providing and using energy are developed.

The Caspian region has the potential to make a significant contribution to ensuring energy security in the rest of the world, and to increasing the diversity of global oil and gas supplies. The Caspian region contains substantial resources of both oil and natural gas, which could underpin a sizeable increase in production and exports over the next two decades. But potential barriers to the development of these resources, notably the complexities of financing and building pipelines through several countries, are expected to constrain this expansion to some degree. In the New Policies Scenario, while Russia will remain a purchaser of Caspian gas, there will be greater diversity in Caspian gas trade as the region expands its access to new markets in Europe and China, which is emerging as a major new customer.

In order to take advantage of the potential in Eurasian countries, it will be essential to reinforce cooperative mechanisms and make considerable efforts to restore and develop infrastructure, and to reduce border inefficiency. Eurasian countries should strive to support and guarantee the flow of supply. In the short term, there are growing concerns for investments in existing and new oilfields and transport infrastructure for energy – especially oil and gas – and also diversifying supply and export routes. For this purpose, it is vital that advantage be taken of the opportunities offered by further regional integration.

References

Aalto, 'dan sonraki p BÜYÜK OLACAK (2012). Russia's Energy Policies: National, Interregional and Global Levels, Edward Elgar.

Balmaceda, M. M. (2007). Energy, integration and disintegration in the former USSR, Routledge.

Baybikova, E. (2011). "International and Regional Development Banks in Northern and Central Eurasia: Overview of Activities in 2010", in Evgeny Vinokurov (ed.) EDB Integration Yearbook 2011, Almaty: Eurasian Development Bank, pp. 321-341.

Bayulgen, O. (2010). Foreign Investments and Political Regimes: The Oil Sector in Azerbaijan, Russia, and Norway, Cambridge University Press.

Eurasian Economic Center. At http://www.eurasian-ec.com/index.php?option= com_content&task = view&id= 2& Itemid=7. (15/03/2011).

Ferguson, R. J. (2006). Lecture 1: Eurasia: Super-region or Zone of Conflict?, Bond University, p. 1. At http://www.international_relations.com/wbeurasia/ WBEA-2006-Lec1.doc.(22/02/2011).

Heifetz, B. (2009). "Russian Direct Investments in EurAsEC and Their Role in Energy and Transport Infrastructure Development in Eurasia", in Evgeny Vinokurov (ed.) *EDB Eurasian Integration Yearbook 2009*: 127-141. Almaty: Eurasian Development Bank. At http://www.decc.gov.uk/assets/decc/what% 20we%20do/uk%20energy%20supply/energy%20markets/gas_markets/1_201 00512151109_e_@@_gassecuritysupply.pdf. (21.06.2011).
http://media.rtl.nl/media/actueel/rtlnieuws/2011/228136.pdf (21.06.2011).

IEA (2010a) International Energy Annual. At http://www.eia.gov/countries/ cab.cfm?fips=RS (27.12.2011).

IEA (2010b). Country Analysis Briefs- Azerbaijan, Russia, Kazakhstan. At http://www.eia.doe.gov (29/11/2011).

IEA (2010c). International Energy Statistics. At http://www.eia.gov/countries/ cab.cfm?fips=RS (28.03.2012).

IEA (2010d). "What does the global energy Outlook to 2035 look like?". World Energy Outlook 2010 Factsheet, p. 3. At http://www.worldenergyoutlook. org/docs/weo2010/factsheets.pdf. (28/03/2012).

IEA (2012). Country Analysis Briefs. Turkmenistan. At http://www.eia.doe.gov. (28/03/2012).

Kalyuzhnova, Y. (2008). Economics of the Caspian Oil and Gas Wealth: Companies, Governments, Policies, Palgrave Macmillan.

Kim, N.K. (2011). Globalization and Regional Integration in Europe and Asia, Ashgate publishing.

Klaassen, G., Grübler, A. and Schrattenholzer, L. (1999). Towards New Energy Infrastructures in Eurasia: A Background Paper, *Interim Report IR-99-17*, December.

Kritzinger, L. (2005). Regional Integration: Concepts, Advantages, Disadvantages and Lessons of Experience, May. At www.sarpn.org.za/documents/.../P1416-RI-concepts_May2005.pdf.

Kurmanalieva, E., Ormantayeva, G. (2011). "Key Macroeconomic Indicators of the CIS Countries in 2010", in Evgeny Vinokurov (ed.) EDB Integration Yearbook 2011, Almaty: Eurasian Development Bank: 313-320.

Linn, J. F. (2006a). Eurasian Integration – a Super-continental Opportunity for Russia, Wolfensohn Center for Development, Kommersant November 24. At http://www.brookings.edu/opinions/2006/1124russia_linn.aspx?p=1 (18/02/2011).

Linn, J. F. (2006b). *The New Impetus towards Economic Integration between Europe and Asia*, at The Fourteenth OSCE Economic Forum,Vienna 23-24 January 2006, EF.NGO/1/06.

Linn, J. F. and Tiomkin, D. (2007). "Economic Integration of Eurasia- Opportunities and Challenges of Global Significance", in Anders Aslund and Marek Dabrowski (eds.) Europe After Enlargement: Newyork, Cambridge University Press.

Linn, J. and Tiomkin, D. (2008). "Economic Integration of Eurasia: Opportunities and Challenges of Global Significance", *Asia Europe Journal*, 4(1).

De Lombaerde, P. and Langenhove, L.V. (2005). Indicator of Regional Integration: Conceptual and Methodological Issues, *IIIS Discussion Paper No. 64*, March.

Maqsimchook, N. (2011). "Chronicle of Eurasian Regional Integration 2010", in Evgeny Vinokurov (ed.) EDB Integration Yearbook 2011, Almaty: Eurasian Development Bank: 246-312.

Marzo, M. (2008). *Security of Natural Gas Supply in Spain*, at VII. ASPO Conference, October 20-21 Barcelona 2008. http://www.aspo-spain.org/aspo7/presentations/Marzo-GasSuply-ASPO7.pdf.

Norling, N. and Swanström, N. (2007). "The Virtues and Potential Gains of Continental Trade in Eurasia", *Asian Survey*, 47(3):351-373.

OECD (2004). Corporate Governance in Eurasia: A Comparative Overview, OECD Publishing.

OECD (2011). OECD Investment Policy Reviews: Ukraine 2011, OECD Publishing.

Overland, I.K., Herschman, H., Kendall-Taylor, A. (2009). "The Resource Course and Authoritarianism in the Caspian Petro States", in Indra Kjaernet Overland,

Heidi Herschman, Andrea Kendall-Taylor, (eds.) Caspian Energy Polics: Azerbaijan, Kazakhstan and Turkmenistan, Routledge.
Pandian, S.G. (2005). "Energy Trade as a Confidence Building Measure between India and Pakistan: A Study of the Indo-Iran Pipeline Project", *Contemporary South Asia*, 14(3):307-320.
Svedberg, M. (2007). "Energy in Eurasia: The Dependency Game", *Transition Studies Review*, 14 (1): 195-202.
Yegorov, Y. and Wirl, F. (2009). Ukranian Gas Transit Game, *Zeitschrift für Energiewirtschaft*, 33(2):147-155. At http://www.springerlink.com/content/93x851nt4u71m431/fulltext.pdf.

The Nabucco Project, Energy Supply Security and International Politics*

Gökhan Özkan

Introduction

It is estimated that global oil consumption will increase one percent per year and global natural gas consumption will increase 1.5 percent per year on average so that total global energy demand will increase 45 percent by 2030.[1] Growing energy demand and its economic and political consequences have enhanced the importance of energy supply security. Energy supply security is defined as the continuous and stable supply of energy from reliable sources, in sufficient amounts, at reasonable prices and via reliable means of transport. It has become more important for world politics since economic parameters are now perceived as being just as important as political and military parameters for national security in the 21st century.[2]

The increasing significance of energy supply security in world politics has multiplied the geopolitical importance of countries that are rich in hydrocarbon reserves. After the end of the Cold War and the dissolution of the Soviet Union, many countries in Central Asia and the Caucasus gained independence. Global and regional actors struggled to shape this hydrocarbon-rich region in line with their interests. This struggle was called the "new great game". The main players in this "new great game" are Turkey, Russia, Iran, China, Afghanistan and Pakistan as regional players and the US, the EU and Japan as global players.[3] As energy security and national security became interconnected; the Middle East and Eurasia, where around 70 percent of the proven oil and natural gas reserves of the world are

---

* This chapter is based on the paper that was presented at the International Conference on Eurasian Economies, Istanbul, November 4-5, 2010.

1 International Energy Agency (2009). World energy outlook 2009: Executive summary. Paris: IEA, p. 3.

2 Demiray, M. ve İşcan, İ. H. (2008). Uluslararasi sistemde güvenlik kavraminin değişimi, ekonomik ve jeopolitik arka plani. Dumlupinar Üniversitesi Sosyal Bilimler Dergisi, 21: p. 161.

3 Amineh, P. and Houweling, H. (2005). Caspian energy: Oil and gas resources and the global market. In P. Amineh and H. Housweling (eds.) Central Eurasia in global politics: Conflict, security and development. Leiden: Koninklijke Brill NV, p. 52.

located, became the "chessboard of the new great game".[4] Rivalry in these regions intensified over exploitation and transportation of these resources.

Various projects were developed to transport these resources to international markets. Regional and global actors are trying to realize pipeline projects in line with their national interests. Nabucco, which will transport natural gas from the Middle East and Caspian Region, is one of these projects. This study evaluates the Nabucco Project within the context of energy supply security and international politics.

1. National Power, Energy Supply Security and International Relations

Energy supply security is important for economic and national security since energy is essential for economic growth[5] and cuts in the energy supply can seriously undermine economic development.[6] Energy supply security became a more significant concept especially after the 1973 oil crisis.

Regional and global actors on the international stage are trying to increase their influence on hydrocarbon-rich regions to enhance their energy supply security. They want to have more say in the exploitation of these resources and how these resources are transported to international markets. So, oil and natural gas are not simple commodities that are exchanged in international markets; they are strategic assets.[7]

After the dissolution of the Soviet Union and after the hydrocarbon-rich Central Asian and Caucasian states gained independence, the Caspian Region became a special focus of regional and global actors and the "new great game" started in the region.[8] Brzezinski likened the region to a chessboard, on which regional and global players struggle to protect their geostrategic and economic interests.[9] The geopolitical importance of Eurasia increased along with the transformation process of the international system from a unipolar system into a multipolar system, in which the EU, Russia and China are potential global powers and Turkey is a regional power that enhances its global capabilities.[10]

---

4 Brzezinski, Z. (1997). The grand chessboard: American primacy and its geostrategic imperatives. New York: Basic Books.

5 Gales, et al. (2007). North versus south: Energy transition and energy intensity in Europe over 200 years. European Review of Economic History, 2: 219-253.

6 German, T. (2009). Pipeline politics: Georgia and energy security. Small Wars & Insurgencies, 20 (2): 344-362.

7 Amineh, P. and Houweling, H. (2005). Caspian energy ..., p. 82.

8 Dhaka, A. (2009). The geopolitics of energy security and the response to its challenges by India and Germany. Geopolitics, 14 (2): p. 279.

9 Brzezinski, Z. (1997). The grand chessboard ....

10 Ariboğan, D.Ü. ve Bilgin, M. (2009). Yeni enerji düzeni siyaseti (neopolitik): Jeopolitik'ten enerjeopolitik'e. Uluslararasi İlişkiler, 5 (20) (Özel Sayi: Enerji): 109-132.

Eurasian energy geopolitics is defined as a struggle for influence, power, and hegemony over the oil and natural gas in Central Asia and the Caucasus.[11] The West and Russia are trying to increase their control over the hydrocarbon resources and the export infrastructure in the region, as can be seen in the Nabucco and the South Stream Projects. Since it is estimated that Azeri and Turkmen natural gas will not be sufficient for both projects, the West and Russia are trying to sign agreements with Azerbaijan and Turkmenistan to fill the Nabucco and South Stream Pipelines.[12]

The Middle East and Eurasia are the richest regions in the world in terms of natural gas and oil, as can be seen from Table 1. These regions are home to almost seventy percent of the world's oil and natural gas reserves. Turkey's geographic location makes it a natural energy bridge between the Middle East, Eurasia and the European Union countries, some of which have the highest per capita energy consumption levels in the world.[13] Turkey is aware of the geostrategic importance of its location and uses its advantage to be an energy hub.[14]

Table 1: Proven Oil and Natural Gas Reserves

Region/Source	Oil (billions of barrels) BP	Oil (billions of barrels) Oil and Gas Journal	Natural Gas (tcf) BP	Natural Gas (tcf) CEDIGAZ	Natural Gas (tcf) Oil and Gas Journal
Eurasia	128.15	98.89	1,890.89	1,900.27	1,993.80
Middle East	755.33	746.00	2,585.35	2,609.32	2,591.65
Europe	15.57	16.66	207.65	218,13	169.09
Africa	117.48	117.06	514.92	514.33	494.08
Asia and Oceania	40.85	34.01	510.69	531.81	430.41
North America	70.31	209.91	308.29	308.46	308.79
C. & S. America	111.21	122.69	272.84	260.10	266.54
World	1,238.89	1,345.21	6,290.64	6,342.41	6,254.36

Source: U.S. Energy Information Administration, 2009.

11 Edwards, M. (2003). The new great game and the new great gamers: Disciples of Kipling and Mackinder. Central Asian Survey, 22 (1): p. 83.

12 German, T. (2009). Pipeline politics ..., p. 345.

13 Bozhilova, D. (2009). Energy security and regional cooperation in south-east Europe. Journal of Balkan and Near Eastern Studies, 11 (3): 293-311.

14 Saiverts, C.R. (2009). Tangled pipelines: Turkey's role in energy export plans. Turkish Studies, 10 (1): 95-108.

The EU countries, because of their growing natural gas needs, became dependent on Russia. Some EU countries are nearly 100 percent dependent on Russian gas. Because of this, the EU is taking steps to diversify its energy supplier countries and pipeline routes to increase its energy supply security.[15] It is supporting projects that will transport natural gas and oil from the Middle Eastern and Caspian regions to the EU without having to pass through Russia. These projects include the Baku-Tiflis-Ceyhan (BTC), Baku-Tiflis-Erzurum (BTE) and Nabucco pipelines. The next section evaluates the Nabucco Project.

2. The Nabucco Project

The main objective of the Nabucco Project is to transport natural gas from the Caspian Region and the Middle East to Europe via Turkey and the Balkans.[16] When completed, the pipeline will be 3300 km long, with a capacity of 30 billion m^3 per year. The Nabucco Pipeline will pass through the territories of Turkey, Bulgaria, Romania, Hungary and Austria. Its estimated cost is around 7.9 billion dollars.

Since the Nabucco Pipeline will decrease the EU's energy dependence on Russia, it is being supported by the EU and the US. According to Mihail Delyagin, President of the Russian Global Issues Institute, the Nabucco Project is the US's geopolitical project. Delyagin argues that the US broke the strategic relationship between Caucasia and Russia through the BTC Pipeline and may be aiming to do the same with Turkmenistan, Uzbekistan and Kazakhstan through the Nabucco Project.[17]

The Nabucco Project is an important pillar of the EU's energy policy. If realized, it will be an important step for the EU towards diversifying supplier countries and energy routes. The EU provided funding for the Nabucco Project for the first time at the beginning of 2009. The Nabucco Intergovernmental Agreement was signed by Turkey and four European countries (Austria, Hungary, Romania, and Bulgaria). When the Nabucco Pipeline is completed, the EU will be able to draw natural gas from the Caspian Region and bypass Russian territory, thus reducing its dependence on Russia.

The Nabucco Project is a project that will enhance Turkey's geopolitical importance. When completed, Turkey will be one of the key countries in the East-

---

15 Yergin, D. (2006). Ensuring energy security. Foreign Affairs, 85 (2): p. 71.

16 Roberts, J. M. (2006). The Black Sea and European energy security. Southeast European and Black Sea Studies, 6 (2): p. 217.

17 Özbay, F. (2009). Nabucco'nun Fendi Güney Akim'i Yendi mi? http://www.bilgesam.org/tr/index.php?view=article&catid=131%3Aenerji&id=398%3Anabucconun-fendi-guney-akimi-yendi-mi&format=pdf&option=com_content&Itemid=146 (Erişim Tarihi 7.4.2010).

West energy corridor. The Nabucco Project is an important project for Turkey also for its own energy security. Turkey wants to diversify its energy supplier countries and routes in the wake of cuts in natural gas supplies from Iran and Russia.[18]

The Nabucco Project is seen as one of the main pillars of the "new great game". The perspectives of regional and global players in respect of the Project differ in line with their national interests as will be examined in the following sections.

2.1 Russia

Russia, the richest country in the world in terms of natural gas reserves, has 26.6 percent of all proven natural gas reserves. Russia is also one of the richest countries in terms of coal and oil reserves. It is the third richest country in terms of coal reserves and the eighth richest country in terms of oil reserves. Russia's natural resources make it one of the most important players in the global energy market.

As oil and natural gas demand has risen in the last 10-20 years as a consequence of the fast economic growth in developing countries such as China and India, as energy supply security has become a national security priority for both developed and developing countries, and as social and political instabilities in the Middle East have increased uncertainties about energy supply from this region, so Russia's importance and leverage in the global energy market have grown. Russia used this leverage to increase its declining influence after the Cold War and became an energy superpower.[19]

Because of its proximity to international markets, Russia also has the advantage of location. Because of its proximity to Europe, 75 percent of its energy exports go to European countries. Russia also wants to exert its influence on the European energy infrastructure. Gazprom, which is majority owned by the Russian state, is trying to establish partnerships with European energy companies. Some argue that Gazprom is acquiring foreign energy companies to support the Kremlin's foreign policy.[20] Russia's policy was criticized by some European countries, which blocked the acquisition of their national energy companies by Russia.

It is argued that Russia is trying to monopolize the transportation, not just of its own hydrocarbon resources but also of the hydrocarbon resources in the Caspian

18 Martin, L. G. (2009). Turkey and Gulf Cooperation Council security. Turkish Studies, 10 (1): 75-93.

19 Sevastyanov, S. (2008). The more assertive and pragmatic new energy policy in Putin's Russia: Security implications for Northeast Asia. East Asia, 25: 35-55.

20 Heinrich, A. (2008). Under the Kremlin's thumb: Does increased state control in the Russian gas sector endanger European energy security? Europe-Asia Studies, 60 (9): p. 1570.

Region, by using its geostrategic location.[21] To do this, it is trying to increase its influence on Central Asian and Caucasian countries in line with its "Near Abroad" policy. It is opposing Western countries' efforts to bypass Russia by linking the supplier countries in the Caspian Region to Europe via Azerbaijan and Turkey.[22]

The military clashes between Georgia and Russia in the summer of 2008 are seen as a reflection of Russia's security and energy policy. The clashes, which changed the balance of power in the Caucasus[23], are evaluated as reflecting the different security perceptions of the West and Russia after the end of the Cold War.[24] Pabst argues that the global financial crisis moved the balance of power towards the East, and that the clashes of August 2008 accelerated this movement.[25]

The clashes between Russia and Georgia may negatively affect pipeline projects supported by the US, such as the Nabucco Project.[26] It is argued that Russia's objective was to take control of the strategic energy export infrastructure rather than to protect its citizens in South Ossetia.[27] The US authorities criticized the clashes, stating that they perceived them as Russia's efforts to divide Georgia and undermine the South Energy Corridor.[28] A report prepared by the European Commission states that incidents in Georgia show that these are critical times for energy security, and that the EU should intensify its efforts with regards to energy supply security.[29]

In opposition to the Nabucco Project, which will transport Caspian natural gas to Europe without passing through Russian territory, Russia developed the

21 Blank, S. (2008). What comes after the Russo-Georgian war? What's at stake in the CIS. American Foreign Policy Interests, 30 (6): 379-391.

22 Rywkin, M. (2008). Russia's place in the world. American Foreign Policy Interests, 30 (5): p. 311.

23 Ekşi, M. (2009). Türkiye-Azerbaycan ilişkileri: Söylemden reelpolitiğe. Avrasya Etüdleri, 36: p. 110.

24 Alexandrova-Arbatova, N. (2009). The impact of the Caucasus crisis on regional and European security. Southeast European and Black Sea Studies, 9 (3): p. 287.

25 Pabst, A. (2009). Central Eurasia in the emerging global balance of power. American Foreign Policy Interests, 31 (3): p. 166.

26 Cheterian, V. (2009). The august 2008 war in Georgia: From ethnic conflict to border wars. Central Asian Survey, 28 (2): p. 164.

27 German, T. (2009). Pipeline politics ..., p. 344.

28 Fried, D. (2008). US-Russia relations in the aftermath of the Georgia crisis. Testimony of Daniel Fried, Assistant Secretary of State for European and Eurasian Affairs, before the House Committee on Foreign Affairs. Washington DC, 9 September 2008. http://www.state.gov/p/eur/rls/rm/109363.htm (Erişim tarihi 25.08.2010).

29 Commission of the European Communities (2008). Green paper: Towards a secure, sustainable and competitive European energy network. Brussels: CEC.

alternative South Stream Project,[30] which aims to transport natural gas from the Caspian Region to Europe via Russia and the Black Sea, bypassing Azerbaijan and Turkey.

The South Stream Pipeline is 3200 km long, 900 km of which will go under the Black Sea. The pipeline passes through Russia-Bulgaria-Romania-Hungary-Austria. It is estimated that the pipeline, which will transport 30 billion m^3 of natural gas per year, will cost around 10 billion euro.

2.2 The United States

The US, which was the only superpower left in the international system after the end of the Cold War, wants to increase its influence over energy resources to sustain its superpower status. The discovery of oil and natural gas reserves in Central Asia and the Caucasus caught the attention of the US together with other regional and global actors, and the US defined the region as strategically vital. The US tried to protect its national interests along with the interests of American energy companies.[31] Exploitation of the Caspian Region energy resources and the critical energy infrastructure to transport these resources became the focus of US's security policy in the region.[32]

In the post-Cold War period, control of the Eurasian Heartland (as defined by Mackinder) became one of the priorities of US foreign policy.[33] It is argued that the US's objective is not only to control the energy resources in the region but also to contain its potential rivals such as China and Russia.[34] China's increasing influence in Central Asia is evaluated as one of the issues that the US must be careful about.[35]

The US's strategy in the region is to support projects that will transport oil and natural gas from the Caspian Region via routes that bypass Russian territories. In doing so, the US is trying to diminish Russia's influence in the region.[36]

30 Larrabee, F.S. (2009). The United States and security in the Black Sea Region. Southeast European and Black Sea Studies, 9 (3): p. 305.

31 Tsantoulis, Y. (2009). Geopolitics, (sub)regionalism, discourse and a troubled 'power triangle' in the Black Sea. Southeast European and Black Sea Studies, 9 (3): p. 249.

32 McDermott, R. N. (2008). United States and NATO military cooperation with Kazakhstan: The need for a new approach. The Journal of Slavic Military Studies, 21 (4): p. 616.

33 Labban, M. (2009). The struggle for the heartland: Hybrid geopolitics in the Transcaspian. Geopolitics, 14 (1): p. 2.

34 İşeri, E. (2009). The US grand strategy and the Eurasian heartland in the twenty-first century. Geopolitics, 14 (1): p. 27.

35 Rywkin, M. (2008). Russia's place in the world. American Foreign Policy Interests, 30 (5): p. 312.

36 Larrabee, F.S. (2009). The United States and Security in the Black Sea Region. Southeast European and Black Sea Studies, 9 (3): p. 304.

The US also wants to prevent Iran from becoming a transit country. It is opposed to Iran's participation in pipeline projects such as the Nabucco Project.

The optimal pipeline route for the US's strategy is the Azerbaijan-Georgia-Turkey-Europe route. Because of this, the US has been the strongest supporter of the BTC, BTE and Nabucco Projects.

2.3 Iran

Iran has the richest natural gas reserves in the world after Russia and the richest oil reserves after Saudi Arabia. It is argued that Iran's importance in global energy geopolitics may increase rapidly if it can take effective advantage of its natural resources.[37] Nonetheless, because it is technologically backward and in dispute with Western countries, it faces many obstacles to exploiting and transporting these resources. Disagreement between Iran, the US and the EU about nuclear energy is hindering cooperation among these countries with regards to the exploitation and transportation of Iran's natural resources. Iran could not be a part of the Nabucco Project because of US political reservations.

Turkey signed the "Natural Gas Memorandum of Understanding (MoU)" with Iran in 2008 to develop its relations with its neighbour and to diminish its energy dependence on Russia. According to the MoU, a 2000 km pipeline from Iran's Assaluyeh region to the Turkish border is going to be built and connected to the pipeline on the Turkish side. The natural gas that will be transported to Turkey through this pipeline will be produced at the South Pars natural gas field in Assaluyeh. Turkey will participate in development projects affecting the 21st, 22nd and 23rd phases of the South Pars Field, where 7.8 percent of the world's natural gas reserves are located. Turkey Petroleum Corporation (TPAO) will exploit these fields. The MoU was extended in 2009. The two countries also agreed to exchange the fields that were allocated to Turkey with richer fields.

The MoU is important for Turkey in many respects. If energy cooperation between Turkey and Iran increases, Turkey's energy dependence on Russia, which is currently around 65 percent, will decrease. Turkey will diversify its energy supplier countries and routes. Trade and bilateral relations between the two countries will develop.

The MoU is important especially with regards to the Nabucco Project. If the proposed pipelines are built, it will be possible to transport Iranian gas to Europe through the Nabucco Pipeline. It will also be possible to transport Turkmen gas across Iranian territory to Turkey. At present, Turkmen gas cannot be included in

---

37 Coşkun, B.B. (2009). Küresel enerji jeopolitiği ve İran. Uluslararasi İlişkiler, 5 (20) (Özel Sayi: Enerji): 179-201.

the Nabucco Project because of uncertainties about the legal status of the Caspian Sea. It is estimated that negotiations about the legal status of the Caspian Sea can move forward if Turkmen gas is included in the Nabucco Project.

2.4 China

One of the most important players of the new great game in Central Asia and the Caucasus is China, whose economic growth rate has surpassed eight percent on average over the last two decades. China invests all over the world to provide the energy it needs for its fast growing economy.

China is trying to play the role of responsible shareholder in the international system, as it did during the Georgia crisis in August 2008. At the same time, it is trying to sustain its strategic partnership with Russia and follow a balanced policy.[38]

According to International Energy Agency forecasts, China's oil demand will surpass 10 million barrels per day by 2020. Since China's resources are not sufficient to meet its own demand, it is forecast that China will import 80 percent of its energy needs. The Caspian region, because of its proximity, is seen as a good alternative for providing the energy China needs. Because of this, China has made big investments in the region.[39]

China has developed various projects to exploit and transport the rich natural resources of the region. The Turkmenistan-Uzbekistan-Kazakhstan-China Natural Gas Pipeline Project, which was developed to transport Turkmen gas to China, has been realized and the first Turkmen gas reached Beijing in February 2010. Transportation of Uzbek and Kazakh gas on this pipeline is also being considered.

China is also investing in the Middle East to meet its energy demand. In 2004, China signed a 100 billion dollar agreement with Iran. According to this agreement, the Chinese state energy firm Sinopec undertook to buy 250 million m^3 LNG from Iran over 25 years and to develop the Yadavaran Field.[40] It is argued that, in light of US preponderance on the high seas, China is developing a project that will link Iranian natural gas and oil to the China-Kazakhstan Pipeline to increase its energy security.

China's interest in oil and natural gas from the Caspian region is welcomed by the countries there since they want to diversify their export markets. Nonethe-

38 Feng, Z. (2008). The Russia-Georgia military conflict: For China, does it mean scrambling between a "strategic partnership" and being a "responsible stakeholder"? Korean Journal of Defense Analysis, 20 (4): p. 316.

39 Ong, R. (2009). China's strategic convergence with Russia. Korean Journal of Defense Analysis, 21 (3): p. 319.

40 Pham, J. p. (2009). China's "surge" in the Middle East and its implications for U.S. interests. American Foreign Policy Interests, 31 (3): p. 181.

less, it is argued that China's interest may create competition among the Nabucco, Turkmenistan-Uzbekistan-Kazakhstan-China and Turkmenistan-Kazakhstan-Russia Pipelines.

2.5 The European Union

Since Europe is a relatively poor region in terms of natural gas reserves and since the EU has to import most of the natural gas it needs, energy security has become a priority for the EU. The EU, as of 2009, is second to the US in terms of natural gas consumption with 471.2 billion m^3 per year. It imports around 50 percent of the natural gas it needs and around 80 percent of this natural gas comes from three countries; Russia, Norway and Algeria. It is estimated that the EU's demand for oil and natural gas will increase since the oil and natural gas resources in the North Sea are declining. Also, nuclear energy is being used less, and coal production is declining because of increasing labour costs. This means that the EU's energy dependence on Russia, Norway and Algeria will increase if it does not diversify its energy resources and supplier countries.

The natural gas and oil reserves in Central Asia and the Caucasus are a possible alternative for the EU, which is trying to diversify its supplier countries and diminish its energy dependence on Russia. It is argued that the EU's regional policies depend on setting up an energy export infrastructure that bypasses Russia with the support of the US.[41]

After the supply of natural gas from Russia to Europe was cut in 2006 in the wake of price disagreements between Russia and the Ukraine and after this problem recurred in subsequent years, the EU decided to intensify its efforts to enhance its energy security [42] The EU countries increased their support for the Nabucco Project and the Nabucco Agreement was signed in July 2009. It is argued that the military clashes between Russia and Georgia in 2008 and a recurrence of the energy crisis between Russia and the Ukraine in 2009 increased the momentum driving the Nabucco Project.

After the energy crisis between Russia and the Ukraine in 2009, the EU declared that energy security had become a priority. The EU Commission drafted the Energy Security and Solidarity Action Plan, according to which the EU plans to follow a common energy policy and increase its bargaining power.

Nonetheless, as bilateral agreements governing the South Stream Project between Russia and the member countries demonstrate, the EU is not proving suc-

---

41 German, T. (2009). Pipeline politics: Georgia and energy security. Small Wars & Insurgencies, 20 (2): p. 345.

42 Saiverts, C. R. (2009). Tangled pipelines ..., p. 95.

cessful at following a synchronized energy policy. Germany has developed the North Europe Natural Gas Pipeline Project with Russia to bypass other countries and buy natural gas directly from Russia. If this project is realized, Germany's energy dependence on Russia will increase and it will be harder for the EU to follow a common energy policy.

It is argued that Russia is trying to create competition among the EU member countries by raising their hopes about becoming an energy hub.[43] Gazprom is trying to enter, not only the Central and Eastern European energy market, but also the French and Italian markets by means of asset swaps and distribution agreements. [44] It is also offering European companies long-term agreements to supply gas directly to the European consumers. Russia's policies have prompted the EU to take steps to diminish the EU's energy dependence on Russia. As a result, EU support for the Nabucco Project has increased.

3. The Nabucco Project and Turkey

The geostrategic location of Turkey between the Caspian region, the Middle East and the EU makes Turkey one of the key countries in the East-West energy corridor and one of the most important players in the new great game in the region.

Turkey became a key country with respect to the transportation of hydrocarbon resources from the Middle East and Caspian region to Europe via routes that bypass Russia after the EU decided to diversify its energy resources and routes. Competition between Turkey and Russia intensified as can be seen by Russia's opposition to the BTC Project.[45]

Despite competing in the field of energy, relations between Turkey and Russia have developed at other dimensions in recent years. Turkey and Russia are cooperating on issues such as international terrorism, Afghanistan, and drug trafficking. Economic relations between the two countries are also progressing rapidly. Turkey imports around two thirds of its natural gas needs from Russia. Turkish businesses are investing in Russia especially in the construction sector, and hundreds of thousands of Russian tourists come to Turkey every year. The trade volume between the two countries amounts to 20 billion dollars. Russia has become Turkey's second largest trade partner.

---

43 Strategic Comments (2009). The Russian-Ukrainian gas dispute. 15 (1), p. 2.

44 Winrow, G.M. (2008). Energy security in the broader Mediterranean. European Security, 17 (1): p. 176.

45 Öniş, Z. and Yilmaz, Ş. (2009). Between Europeanization and Euro-Asianism: Foreign policy activism in Turkey during the AKP era. Turkish Studies, 10 (1): p. 10.

Despite Russian-Turkish cooperation in various areas, the interests of the two countries do sometimes clash over control of the strategic energy infrastructure. It is argued that the Nabucco Project is an alternative to the South Stream Project and there is not enough natural gas in the region to realize both projects. The Nabucco Project may diminish Russia's role in transporting hydrocarbon resources from the Caspian Region to Europe, as did the BTC.

At present, the BTC Pipeline plays a very important role in transporting oil from the Caspian Region to Western markets. The BTC Pipeline, which transports only Azeri oil, is expected to also transport Kazakh oil after realization of the Trans-Caspian Project. Once the Kerkuk-Yumurtalik Oil Pipeline is running at full capacity, the amount of oil that will be supplied to the international markets from Ceyhan will surpass two million barrels a day and Ceyhan will become an important energy terminal.

The Nabucco Project may enhance Turkey's role in the East-West corridor even further. If the Nabucco Project is realized, Turkey will be able to act as the energy bridge, transporting natural gas to Europe, not only from the Caspian Region but also from the Middle East and Africa.

Not only do the Nabucco, BTC and BTE pipelines turn Turkey into an energy corridor and increase its geopolitical importance, they also increase its energy supply security. Like the EU, Turkey is energy dependent on other countries and its dependence is expected to increase, which is why Turkey is also trying to diversify its energy resources, energy supplier countries and routes.

It is argued that if Turkey becomes one of the key countries in the East-West corridor and if it becomes a strategic country for EU energy supply security, Turkey's negotiations to join the EU may proceed more positively.[46] The BTC pipeline has improved relations between the EU and Turkey since it has reduced EU energy dependence on Russia. It is argued that if the Nabucco Project is realized, Turkey's geopolitical importance for the EU will increase and EU-Turkish relations will develop faster.

Conclusion

Turkey's unique geostrategic location makes it simultaneously a Middle Eastern, European and Eurasian country and increases its potential to be one of the key countries in the East-West and North-South energy corridors. In light of the strategies being adopted by the regional and global players in the new great

46 Tekin, A. and Williams, P. A. (2009). EU-Russian relations and Turkey's role as an Energy corridor. Europe-Asia Studies, 61 (2): 337-356.

game, Turkey should pursue proactive rather than reactive policies. It can become one of the key countries of the East-West and North-South energy corridors if it can take advantage of its geostrategic location in the Middle East and the Caspian region, which is home to around 70 percent of all proven natural gas and oil reserves in the world.

Nabucco is possibly the most important project in the East-West energy corridor. If it is realized, Turkey will become one of the key countries for transportation of natural gas, not only from Central Asia and the Caucasus, but also from the Middle East and Africa. Since energy security has become a strategic priority for the EU, Turkey's geostrategic importance for the EU will increase and Turkey's negotiations to join the EU will proceed more positively. Turkey will be able to host the "alliance of civilizations" at the energy dimension.

References

Amineh, P. and Houweling, H. (2005). Caspian energy: Oil and gas resources and the global market. In P. Amineh and H. Housweling (eds.) Central Eurasia in global politics: Conflict, security and development. Leiden: Koninklijke Brill NV.

Alexandrova-Arbatova, N. (2009). The impact of the Caucasus crisis on regional and European security. Southeast European and Black Sea Studies, 9 (3): 287-300.

Ariboğan, D.Ü. ve Bilgin, M. (2009). Yeni enerji düzeni siyaseti (neopolitik): Jeopolitik'ten enerjeopolitik'e. Uluslararasi İlişkiler, 5 (20) (Özel Sayi: Enerji): 109-132.

Blank, S. (2008). What comes after the Russo-Georgian war? What's at stake in the CIS. American Foreign Policy Interests, 30 (6): 379-391.

Bozhilova, D. (2009). Energy security and regional cooperation in south-east Europe. Journal of Balkan and Near Eastern Studies, 11 (3): 293-311.

Brzezinski, Z. (1997). The grand chessboard: American primacy and its geostrategic imperatives. New York: Basic Books.

Cheterian, V. (2009). The August 2008 war in Georgia: From ethnic conflict to border wars. Central Asian Survey, 28 (2): 155-170.

Commission of the European Communities (2008). Green paper: Towards a secure, sustainable and competitive European energy network. Brussels: CEC.

Coşkun, B.B. (2009). Küresel enerji jeopolitiği ve İran. Uluslararasi İlişkiler, 5 (20) (Özel Sayi: Enerji): 179-201.

Demiray, M. ve İşcan, İ. H. (2008). Uluslararasi sistemde güvenlik kavraminin değişimi ekonomik ve jeopolitik arka plani. Dumlupinar Üniversitesi Sosyal Bilimler Dergisi, 21: 141-171.

Dhaka, A. (2009). The geopolitics of energy security and the response to its challenges by India and Germany. Geopolitics, 14 (2): 278-299.

Edwards, M. (2003). The new great game and the new great gamers: Disciples of Kipling and Mackinder. Central Asian Survey, 22 (1): 83-102.

Ekşi, M. (2009). Türkiye-Azerbaycan ilişkileri: Söylemden reelpolitiğe. Avrasya Etüdleri, 36: 95-112.

Feng, Z. (2008). The Russia-Georgia military conflict: For China, does it mean scrambling between a "strategic partnership" and being a "responsible stakeholder"? Korean Journal of Defense Analysis, 20 (4): 305-318.

Fried, D. (2008). US-Russia relations in the aftermath of the Georgia crisis. Testimony of Daniel Fried, Assistant Secretary of State for European and Eurasian Affairs, before the House Committee on Foreign Affairs. Washington DC, 9 September 2008. http://www.state.gov/p/eur/rls/rm/109363.htm (Erişim tarihi 25.08.2010).

Gales, et al. (2007). North versus south: Energy transition and energy intensity in Europe over 200 years. European Review of Economic History, 2: 219-253.

German, T. (2009). Pipeline politics: Georgia and energy security. Small Wars & Insurgencies, 20 (2): 344-362.

Heinrich, A. (2008). Under the Kremlin's thumb: Does increased state control in the Russian gas sector endanger European energy security? Europe-Asia Studies, 60 (9): 1539-74.

International Energy Agency (2009). World energy outlook 2009: Executive summary. Paris: IEA.

İşeri, E. (2009). The US grand strategy and the Eurasian heartland in the twenty-first century. Geopolitics, 14 (1): 26-46.

Labban, M. (2009). The struggle for the heartland: Hybrid geopolitics in the Transcaspian. Geopolitics, 14 (1): 1-25.

Larrabee, F.S. (2009). The United States and Security in the Black Sea Region. Southeast European and Black Sea Studies, 9 (3): 301-315.

Martin, L.G. (2009). Turkey and Gulf Cooperation Council security. Turkish Studies, 10 (1): 75-93.

McDermott, R.N. (2008). United States and NATO military cooperation with Kazakhstan: The need for a new approach. The Journal of Slavic Military Studies, 21 (4): 615-641.

Ong, R. (2009). China's strategic convergence with Russia. Korean Journal of Defense Analysis, 21 (3): 315-328.

Öniş, Z. and Yilmaz, Ş. (2009). Between Europeanization and Euro-Asianism: Foreign policy activism in Turkey during the AKP era. Turkish Studies, 10 (1): 7-24.
Özbay, F. (2009). Nabucco'nun Fendi Güney Akim'i Yendi mi? http://www. bilgesam.org/tr/index.php?view=article&catid=131%3Aenerji&id=398%3Anabucconun-fendi-guney-akimi-yendi-mi&format=pdf&option=com_content&Itemid=146 (Erişim Tarihi 7.4.2010)
Pabst, A. (2009). Central Eurasia in the emerging global balance of power. American Foreign Policy Interests, 31 (3): 166-176.
Pham, J.P. (2009). China's "surge" in the Middle East and its implications for U.S. interests. American Foreign Policy Interests, 31 (3): 177-193.
Roberts, J.M. (2006). The Black Sea and European energy security. Southeast European and Black Sea Studies, 6 (2): 207-223.
Rywkin, M. (2008). Russia's place in the world. American Foreign Policy Interests, 30 (5): 310-314.
Saiverts, C.R. (2009). Tangled pipelines: Turkey's role in energy export plans. Turkish Studies, 10 (1): 95-108.
Sevastyanov, S. (2008). The more assertive and pragmatic new energy policy in Putin's Russia: Security implications for Northeast Asia. East Asia, 25: 35-55.
Strategic Comments (2009). The Russian-Ukrainian gas dispute. 15 (1), p. 2.
Tekin, A. and Williams, P. A. (2009). EU-Russian relations and Turkey's role as an Energy corridor. Europe-Asia Studies, 61 (2): 337-356.
Tsantoulis, Y. (2009). Geopolitics, (sub)regionalism, discourse and a troubled 'power triangle' in the Black Sea. Southeast European and Black Sea Studies, 9 (3): 225-241.
Winrow, G.M. (2008). Energy security in the broader Mediterranean. European Security, 17 (1): 161-183.
Yergin, D. (2006). Ensuring energy security. Foreign Affairs, 85 (2): 69-82.

Current Condition of OPEC as the Main Energy Player, and Position of OPEC and Particularly Saudi Arabia in the Face of Increasing Global Energy Demand

Müslüme Narin

Introduction

In the past 50-60 years, the world has witnessed considerable development and progress in fields such as standard of living, health, life expectancy and welfare. These achievements have contributed to the availability of abundant and cheap sources of energy, mainly in the form of crude oil from the Middle East. Adequate global energy supplies are essential to both individual countries and the world as a whole for sustainable development, proper functioning of the economy, and human welfare. Thus, the availability of energy in the forms and quantities required by the economy and societies must be ensured and secured.

Compared with today, energy was easily obtained until 1970. However after the restriction of oil production by OPEC in 1973, the world was hit by oil crises and began to face frequent rises in oil prices. This led to negative influences on the world economies, such as rising inflation, reduction of production, and shortage of commodities and services.

1. Increase in World Primary Energy Demand (Current Situation and Future)

Population growth, expanding industrialization, and improving lifestyles, principally in developing countries, are the main causes of the continuous increase in world energy demand.

1.1 World Population and World Gross Domestic Product Growth

The world population reached 6.852 billion in 2010 and is increasing by 76 million per year. Between 1950 and 2010, the world population tripled. When looked at region by region, and in some countries, it can be seen that a large part of the world population lived in less developed countries as of 2010. The population in less developed countries reached 4.318 billion in 2010. And, while the world's population was 6.852 billion in 2010, it will be 8.249 billion in 2030 (8.536 billion in 2035).[1]

Strong long-term growth in gross domestic product (GDP's) in the emerging economies of non-OECD countries increases their growing energy demand (Table 1).

Table 1: GDP Growth (Annual percent change)

	1993-2002	2003	2004	2005	2006	2007	2008	2009	2010	Projections 2011	2012	2016
World	3.3	3.6	4.9	4.6	5.2	5.4	2.9	-0.5	5.0	4.4	4.5	4.7
Advanced Economies	2.8	1.9	3.1	2.7	3.0	2.7	0.2	-3.4	3.0	2.4	2.6	2.4
United Stated	3.4	2.5	3.6	3.1	2.7	1.9	0.0	-2.6	2.8	2.8	2.9	2.7
Euro Area	2.1	0.8	2.2	1.7	3.1	2.9	0.4	-4.1	1.7	1.6	1.8	1.7
Japan	0.8	1.4	2.7	1.9	2.0	2.4	-1.2	-6.3	3.9	1.4	2.1	1.2
Other Advanced Economies	3.8	2.6	4.1	3.4	3.9	4.0	1.1	-2.3	4.3	3.2	3.3	3.1
Emerging and Developing Economics	4.1	6.2	7.5	7.3	8.2	8.8	6.1	2.7	7.3	6.5	6.5	6.8
European Union	2.4	1.5	2.6	2.2	3.5	3.2	0.7	-4.1	1.8	1.8	2.1	2.1

Sources: IMF. International Monetary Fund. World Economic and Financial Surveys World Economic Outlook (WEO). Tensions from the Two-Speed Recovery: Unemployment, Commodities, and Capital Flows. April 2011. p. 181.

In all non-OECD regions combined, economic activity, as measured by GDP in purchasing power parity terms, increases by 4.4 percent per year on average, compared with an average of 2.0 percent per year for OECD countries (Figure 1).

1 U.S. Census Bureau. World Population Summary. 2011.

Fig. 1: World gross domestic product (GDP) by purchasing power parity, reference case, 2005-2035 (billion dollars, 2005)

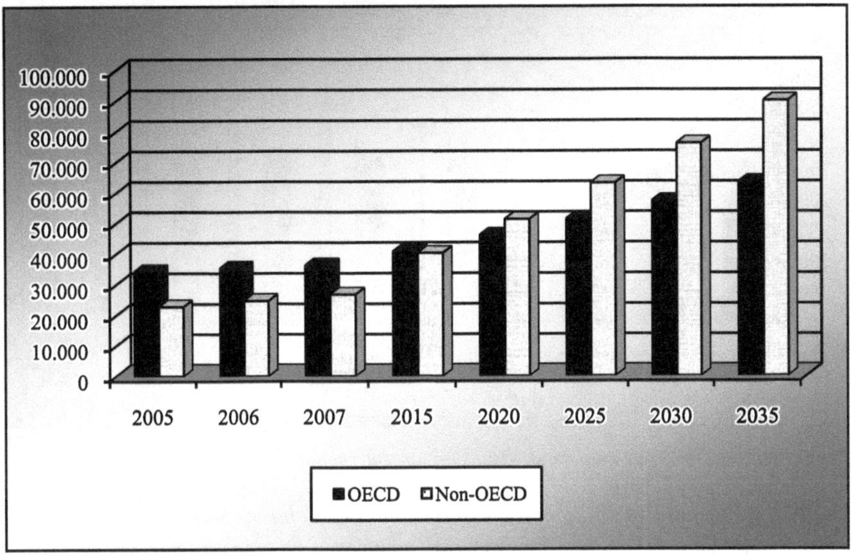

Source: EIA. U.S. Energy Information Administration, International Energy Outlook 2010.

1.2 World Total Energy Consumption

Population and economic growth in developing countries will both increase world demand for energy, especially fossil fuels. While demand for energy is likely to increase less in a number of developed countries, this demand will likely be offset by demand growth in the developing countries. Projections of world energy demand suggest that demand will increase by 49% in the period 2007-2035 in the reference case; while total energy demand in non-OECD countries will increase by 84% (Figure 2).

Fig. 2: World marketed energy consumption, 2007-2035 (quadrillion Btu)

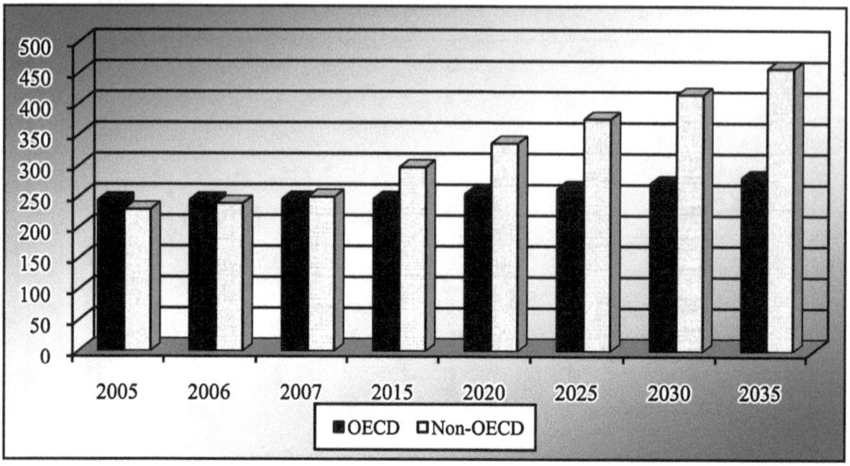

Source: EIA. U.S. Energy Information Administration, International Energy Outlook 2010.

Table 2: World total energy consumption by fuel, reference case, 2005-2035

World (Quadrillion Btu)	History			Projections					Average annual percent change, 2007-2035
	2005	2006	2007	2015	2020	2025	2030	2035	
Liquids	170.4	172.8	174.7	179.3	186.0	197.2	210.0	223.6	0.9
Natural Gas	106.3	108.3	112.1	129.1	141.2	150.2	155.8	162.0	1.3
Coal	122.3	126.4	132.4	139.1	152.4	167.8	185.6	206.3	1.6
Nuclear	27.5	27.8	27.1	32.2	37.4	41.1	43.9	47.1	2.0
Other	46.2	47.9	48.8	63.8	73.4	82.4	91.2	99.8	2.6
Total World	472.7	483.1	495.2	543.5	590.5	638.7	686.5	738.7	1.4

Sources: IEA. International Energy Agency. Key World Energy Statistic 2010.

The IEO2010 reference case projects that the consumption of the marketed energy from all fuel sources will increase over the 2007-2035 projection period. Increases in both the world population and the economic growth will increase global energy demand in general and thus the demand for fossil fuels as well. When compared with other fossil fuels such as natural gas and coal, demand for liquid fuels (oil) will increase less by 2035 (Table 2). Fossil fuels (liquids, natural gas, and coal) are expected to continue supplying much of the energy used worldwide. Although liquid fuels remain the main source of energy, their share in the world's marketed energy consumption is expected to fall in the period

2007-2035 as high oil prices throughout the world may lead many energy users to switch away from liquid fuels when feasible.

1.3 World Liquid Fuel Production and Consumption

Given their importance in the transportation and industrial end-use sectors, liquid fuels remain the world's largest energy source. Worldwide consumption of liquid fuel, which was 84.8 million barrels per day in 2007, is expected to rise to 92.1 million and 110.6 million barrels per day in 2020 and 2035 successively (Figure 3).

Fig. 3: World total liquids production, reference case, 2006-2035

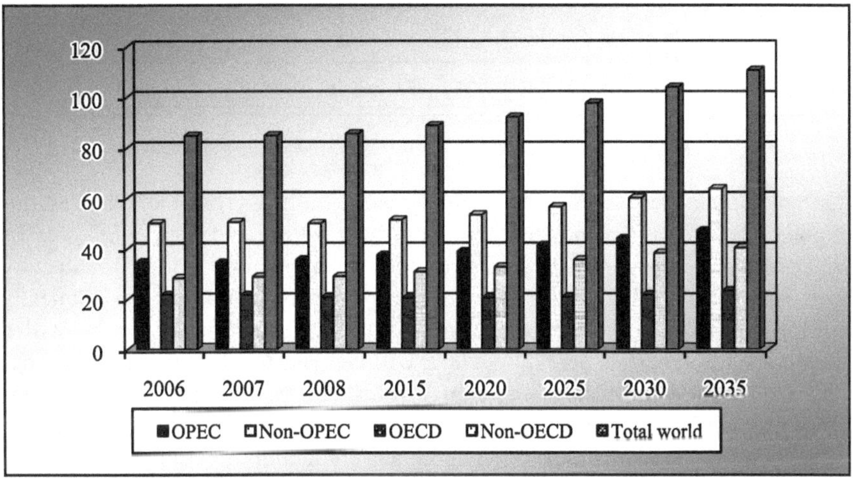

Source: EIA. U.S. Energy Information Administration, International Energy Outlook 2010.

To meet the increase in world demand, liquid fuel production will increase by a total of 25.8 million barrels per day from 2007 to 2035. The IEO2010 assumes that OPEC countries will invest in incremental production capacity in order to maintain a share of approximately 40% of total world liquids production (Figure 4) through 2035. Increasing volumes of conventional liquids from OPEC producers will contribute 11.5 million barrels per day to the total increase in world liquids production, and supplies from non-OPEC countries will add another 4.8 million barrels per day.[2]

2 EIA. U.S. Energy Information Administration, International Energy Outlook 2010. p. 250.

Fig. 4: OPEC and Persian Gulf share of world production, 2006-2035

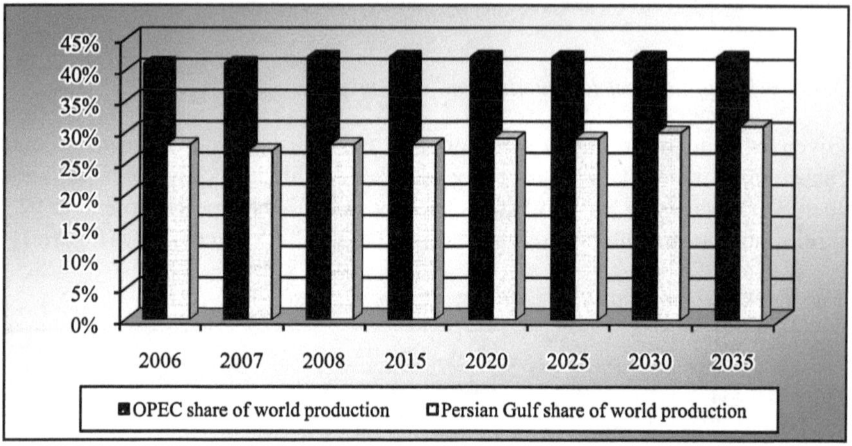

Source: EIA. U.S. Energy Information Administration, International Energy Outlook 2010.

Unconventional resources, such as oil sands, extra-heavy oil, biofuels, coal-to-liquids, gas-to-liquids, and shale oil, from both OPEC and non-OPEC sources will grow on average by 4.9% per year over the projection period. Permanently high oil prices will allow unconventional resources to become economically competitive. World production of unconventional liquid fuels, which totalled only 3.4 million barrels per day in 2007, will increase to 12.9 million barrels per day and account for 12% of the total world liquids supply in 2035. Oil sands from Canada and biofuels, largely from Brazil and the United States, are the largest components of future unconventional production, providing a combined 70% of the increment in total unconventional supply over the projection period.[3]

World liquid fuel consumption, which was 86.1 million barrels per day in 2007, will increase to 92.1 million and 110.3 million barrels per day in 2020 and 2035 successively. While OECD demand will remain about the same, liquid consumption for non-OECD countries will increase by 1.8% per year on average over the projection period (Figure 4).

Although world liquids consumption actually declined in 2008 (to 85.8 million barrels per day) and again in 2009 (to an estimated 84.1 million barrels per day) as the global economic crisis (recession) deepened, it was expected to recover in 2010 and beyond as economic growth resumed. In the long term, world liquids consumption will increase despite rising world oil prices. World oil prices will

3 EIA. U.S. Energy Information Administration, International Energy Outlook 2010. p. 252.

remain above $90 per barrel (in real 2008 dollars) after 2014 and rise to more than $130 per barrel by 2035.[4]

Fig. 4: World total liquids consumption, reference case, 2005-2035

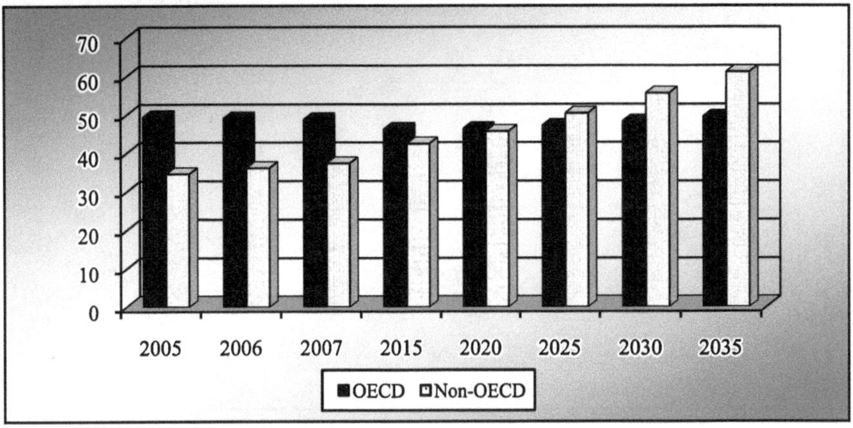

Source: EIA. U.S. Energy Information Administration. 2010. International Energy Outlook 2010. p. 150.

1.4 World Oil Prices

The impact of world oil prices on energy demand is a considerable source of uncertainty in the IEO2010 projections. In the reference case, the world oil price increases from $59 per barrel in 2009 to $70 per barrel in 2010 and then rises to $95 and $133 in 2015 and 2035 ($224 per barrel in nominal terms) successively (Figure 5).

Average oil prices increased strongly from 2003 to mid-July 2008, when prices collapsed as a result of concerns about the deepening recession. In 2009, oil prices trended upward throughout the year, from about $42 per barrel in January to $74 per barrel in December. Oil prices have been especially sensitive to demand expectations, with producers, consumers, and traders continually looking for an indication of possible recovery in world economic growth and a likely corresponding increase in oil demand. On the supply side, OPEC's above-average compliance with agreed production targets increased the group's spare capacity to roughly 5 million barrels per day in 2009. Further, many of the non-OPEC projects that were delayed during the price slump in the second half of 2008 have not yet been revived.

---

4 EIA. U.S. Energy Information Administration. 2010. International Energy Outlook 2010. p. 23.

Fig. 5: World oil prices in three cases, 2008-2035

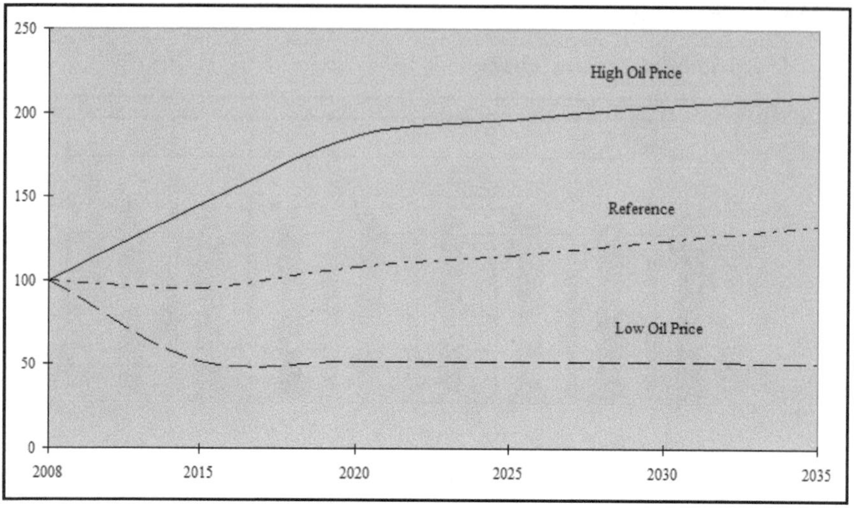

Source: EIA. U.S. Energy Information Administration, International Energy Outlook 2010.

After 2 years of declining demand, world liquids consumption increased in 2010 and strengthened thereafter as the world economies recovered fully from the effects of the recession.[5]

2. The Importance and History of Crude Oil

As an energy resource, crude oil has great importance in human lives. Almost every sector in an economy directly or indirectly depends on oil; therefore, changes in the oil market and oil prices cause various impacts on both national and world economies through chain reactions.

Crude oil accounts for most of the fuel used by people today. Because oil is liquid, it is easy to mine by drilling and pumping rather than excavation, and it is easy to transport in tankers and pipelines. Nevertheless, the history of oil supply has been dominated by the time and place of discoveries, with enormous impacts on the history of the 20th century. It has also been dominated by a few individuals, companies, and nations, with greed, and superb intelligence. Furthermore, petroleum and petroleum products have been minor components of international trade for thousands of years.

5 EIA. U.S. Energy Information Administration. 2010. International Energy Outlook 2010. p. 2.

2.1 History of Crude Oil and Supply

Throughout the history of crude oil, which started well over a hundred years ago, the price of oil has really not fluctuated until recent years. In fact, oil in one form or another has been used since ancient times, and is even more important today in all aspects of life, such as the economy, politics and technology. This rise in importance was mostly due to the invention of the internal combustion engine, the rise in commercial aviation, and the increasing use of plastic.

The modern oil era began in Northwest Pennsylvania in the mid-19th century, as shallow fields were tapped. The early days of the oil industry were characterized by boom and bust, as new discoveries first overwhelmed demand, and then lagged behind it. Crude oil prices fell from $10 a barrel in January 1861 to 10¢ a barrel in December 1861, but were up to $7.25 a barrel again by September 1863, down to $2.40 in 1867.[6]

Since its inception in 1960, OPEC has tried to control the price of oil by restricting the amount that was being produced on a daily basis. Historically, oil prices have been stable, although in recent years the price of oil has rocketed and slumped in lightning fashion.

The Financial Times has used the label the "New Seven Sisters" to describe a group of what it argues are the most influential national oil and gas companies based in countries outside of the OECD. This group of companies consists of China National Petroleum Corporation (China), Gazprom (Russia), National Iranian Oil Company (Iran), Petrobras (Brazil), Petroleos de Venezuela S.A (Venezuela), Petronas (Malaysia), and Saudi Aramco (Saudi Arabia).[7]

2.2 The World's largest Oil Fields and Oil Reserves

Oil is known to have been formed more than half a billion years ago. The oil and gas migrated through interconnected porous rock, and accumulated in pools when stopped by impermeable cap rock. The world's largest oilfields and the time of their formation are shown in Table 3.

6 Live Oil Prices. About Crude Oil. 2011.

7 Hoyos, Carola. 2007. "The new Seven Sisters: oil and gas giants dwarf western rivals". The Financial Times, March 11, 2007.

Table 3: The World's Largest Oilfields

Oilfield	**Location**	**Discovered**	**Started Production**	**Recoverable Oil, Past and Future** (Billion Barrels)	**Production** (Million Barrels/day)
Ghawar	Saudi Arabia	1948	1951	75-83	5
Burgan	Kuwait	1938	1948	66-72	1.7-2
Cantarell	Mexico	1976	1981	35	0.495
Bolivar Coastal	Venezuela	1917		30-32	2.6-3
Safaniya-Khafji	Saudi Arabia/ Neutral Zone	1951		30	1.5
Kashagan	Kazakhstan	2000		30	
Azadegan	Iran	2004		26	0.035
Tengiz	Kazakhstan	1979	1993	15-26	0.53
Sugar Loaf	Brazil	2007		Possibly 25-40	
West Qurna	Iraq	1973		15-21	0.18-0.25
Rumaila	Iraq	1953		17	1.3
Romashkina	Volga-Ural, Russia	1948	1949	16-17	0.301
Samotlor	West Siberia, Russia	1965	1969	14-16	0.844
Majnoon	Iraq	1975		11-20	0.5
Shaybah	Saudi Arabia	1998	1998	15	0.5
Prudhoe Bay	United States, Alaska	1969		13	0.9
Priobskoye	West Siberia, Russia	1982	2000	13	0.680
Lyantorskoye	West Siberia, Russia	1966	1979	13	0.168
Abqaiq	Saudi Arabia	1941		12	0.4
Serir	Libya	1961		12	
Berri	Saudi Arabia	1964		12	
Zakum	Abu Dhabi, UAE	1965	1967	12	
Manifa	Saudi Arabia	1957		11	
Fyodorovskoye	West Siberia, Russia	1971	1974	11	1.9
Ahwaz	Iran	1958		10.1	0.7
Agha Jari	Iran	1937		8.7	0.2
Kirkuk	Iraq	1927	1934	8.5	0.48
Ferdows/Mound/Zagheh	Iran	2003		7-9	
Tupi	Santos Basin Brazil	2007		5-8	
Chicontepec	Mexico	1926		6.5	

Source: Dahl, Carol Ann. 2004. International Energy Markets, Understanding Pricing, Policies, and Profits. Tulsa, Oklahoma: PennWell Press. p. 13; Robelius, Frerdik. 2005. Giant Oil Fields of the World. Presentation AIM Industrial Contact Day, Monday 23rd May 2005; EIA. U.S. Energy Information Administration. 2011. Country Analysis Briefs; Wikipedia, 2011.

As shown in Table 3, the world's largest oilfields that were discovered in the last two decades (since 1990) are Shaybah-Saudi Arabia (1998), Kashagan-Kazakhstan (2000), Ferdows-Iran (2003), Azadegan-Iran (2004), Sugar Loaf-Brazil (2007), and Tupi-Santos Basin-Brazil, proving that new oil reserves are still being discovered.

The world gets its daily ration of 85 million barrels of oil from more than 4,000 fields. Most of these are small – less than 20 thousand barrels per day. Giants, producing more than 100 thousand bpd, account for just 3%. Then there are the megafields that gush out 1 million bpd. Amongst all the fields, these are the most important sources of energy that are worth fighting over. In figuring the top 10 fields of the future, we're not interested in most of the giants of yesteryear, and not necessarily even the giants of today.[8]

1. *Ghawar. Saudi Arabia.* Ghawar, in Saudi Arabia, ranks first on that list. It is thought to have had about 75-83 billion barrels of recoverable oil in place. This field is still going strong, producing 5.7 million bpd from six main producing areas.[9]
2. *West Qurna. Iraq.* West Qurna is one of Iraq's oilfields, located North of Rumaila field, West of Basra. It's believed to hold 43 billion barrels of recoverable reserves, making it the second largest field in the world after Ghawar oilfield. The field has been off limits to Western firms for years due to long sanctions.[10] In 2009, a deal with Exxon and Shell, two of the largest publicly listed oil companies, gave them entry into a country that boasts one of the world's largest reserves of petroleum. In 2010, a joint venture between ExxonMobil and Royal Dutch Shell was awarded the contract to develop the 9 billion barrel first phase of the West Qurna oilfield. They aim to raise output from 300,000 bpd to 2.3 million bpd. It's hard to say these two big oil companies from the countries that invaded Iraq in 2003 are getting a sweetheart deal since the contract stipulates that the government of Iraq retains ownership of the field and the oil.
3. *Majnoon. Iraq.* With massive reserves in a relatively small area near the Euphrates River in Southern Iraq, the field's abundance was so mind-boggling that it was named Majnoon, which means "crazy" in Arabic. This easy oil hasn't been developed, partly because of its location so close to the Iranian border. In the 1980s, during the Iran-Iraq war, managers reportedly buried the wells, concerned that they might be targeted by Iranian forces. The field produces just 0.5 million barrels/day now.
4. *Rumaila. Iraq.* In November 2010, British giant BP and China National Petroleum won the first oil contract of the post-Saddam era to redevelop

8 Helman, Christopher. 2010. The World's Biggest Oil Reserves. Forbes.

9 Greg Croft Inc. The Ghawar Oil Field, Saudi Arabia. 2011.

10 Chon, Gina. Iraq Awards West Qurna-1 Oil Field to Exxon, Shell. The Wall Street Journal. Nowember 5, 2009.

Rumaila. Located on the border with Kuwait, the field is already producing 1.3 million bpd, about half of Iraq's total production.

5. *Khuzestan. Iran.* Not just a field, Khuzestan is the province where 90% of Iran's oil is produced. It borders Iraq and is home to the Ahwaz field, which is believed to produce 0.7 million barrels/day, and the Azadegan field, which is being developed by China's Sinopec under a $70 billion deal made in 2004. Last year, Iran trumpeted the discovery of an 8 billion barrel field in the province called Soussangerd. Iraq's Majnoon field is thought to extend across the border into Khuzestan territory.
6. *Kashagan. Kazakhstan.* Discovered in the Caspian Sea in 2000, Kashagan has recoverable reserves of about 30 billion barrels. The reservoir is deep, the oil is highly corrosive, and getting it is costly. Production is expected to reach 1.5 million bpd by the end of the decade.
7. *Khurais. Saudi Arabia.* In 2009 Saudi Aramco put the finishing touches to the $10 billion development of Khurais. This included pipelines to bring 2 million bpd of seawater for injection under the field. This water injection, a technique perfected by the Saudis at Ghawar, is the key to Khurais and its 1.2 billion bpd of output. The field was partially developed in the 1980s with peak production in the order of 140,000 bpd.
8. *Tupi. Brazil.* Discovered off Rio de Janeiro in 2007, the Tupi field was revolutionary for Brazil. Tupi has established the region as one of the world-class oil and gas basins in the world. The fields are generally under more than one mile of water, three miles of sand and rock, then another mile of solid salt. Because the salt layer scrambles the signals of seismic testing gear, figuring out where to drill the well was extremely difficult and extremely costly.
9. *Carabobo. Venezuela.* The country's oil-rich Carabobo area in North-Central Venezuela boasts huge petroleum reserves of an estimated 15 billion barrels. But it's a heavy grade of crude – thicker, with a higher percentage of impurities such as sulphur – and more difficult to refine than the lighter grades more commonly used around the world.[11] Venezuela established joint ventures between PDVSA (Petroleos de Venezuela S. A.) the state oil company – and two consortia of foreign companies to develop the Carabobo fields in the South American nation's heavy crude belt. Bidding on seven blocks in the Carabobo section of Venezuela's Orinoco heavy oil belt took place at the end of January 2010.
10. *North Slope. Alaska.* By 2025, without new development, North Slope oil output will decline to the point where there won't be enough oil to keep the Trans-Alaskan Pipeline System running. A 2008 Department of Energy study found that the North Slope, including the National Petroleum Reserve parts of the Alaska National Wildlife Refuge and offshore Beaufort and Chukchi Seas could yield nearly 40 billion barrels of oil and more than 125

11 Gelsi, Steve. Venezuela rolls out welcome for foreign oil companies. Carabobo region to draw bidders from Western oil giants. The Wall Street Journal. MarketWatch. January 24, 2010.

trillion cubic feet of natural gas. BP, ExxonMobil and others are willing to build a $30 billion pipeline to carry gas down to the Lower 48 (Table 3).[12]

Table 4: World's Largest Oil and Gas Companies in 2009

Oil Companies	Top Proven World Oil Reserves 2009 Billion Barrels	Rank by 2009 Oil Equivalent Reserves	Crude Oil Production 2009 Million Barrels	Rank by 2009 Crude Oil Production	Refinery Capacity 2009 Thousand bbl/d	Rank by 2009 Refinery Capacity	Petroleum Export 2008, Million bbl/d	Rank by 2009 Export
WORLD	1,322		73.7		85,460			
Saudi Arabian Oil Company (Saudi Arabia) Saudi Aramco	260.1	1	8.3	2	2,080	9	7.3	198
National Iranian Oil Company (Iran)	136.15	3	4.0	4	1.451	15	2.5	195
Iraq National Oil Company (Iraq)	115.0	4	2.4	11	598	31	1.8	188
Kuwait Petroleum Corporation (Kuwait)	101.5	5	2.3	8	889	22	2.1	194
Petroleos de Venezuela S.A. (Venezuela)	99.4	7	2.2	10	1,282	18	1.7	191
United Arab Emirates (UAE) – Abu Dhabi National Oil Company ADNOC) – Dubai Petroleum Company (UAE)	93.8 4.0 Total 97.8	7	2.4	7	781	25	2.3	196
National Oil Company (NOC) (Libya)	44.3	9	1.7	17	378	43	1.5	187
Nigerian National Petroleum Corp. (Nigeria)	36.2	10	2.2	13	505	34	1.9	189
Qatar General Petroleum Corporation (Qatar)	25	14	0.8	21	200	63	1.1	184
Sonatrach (Algeria)	12.2	15	1.7	15	450	39	1.8	190
Petroleos de Ecuador (Ecuador)	6.5	24	0.5	29	176	67	0.3	178

Source: EIA. U.S. Energy Information Administration. 2011. Country Analysis Briefs.

The world's largest oil and gas companies, top proven world oil reserves, production of these companies, refinery capacity, and petroleum export in 2009 are shown in Table 4.

---

12 Helman, Christopher. 2010. The World's Biggest Oil Reserves. Forbes.

2011 in the oil and gas industry brings with it a new set of financial results reflecting the operating prowess of the industry's biggest oil companies. For anyone following the industry, the past year has been anything but boring, whether it was the disaster that unfolded before the world's eyes in the Gulf of Mexico or the fast developing shale gas boom in North America.

2010 saw a number of strategic alliances being formed or strengthened across the energy sector in the name of breaking new technological ground and new geographies. The Middle East got its fair share of attention from the industry giants. Abu Dhabi's massive sour gas project at Shah finally found a partner in the shape of Occidental Petroleum to fill the void left nearly a year ago by ConocoPhillips.

Qatar, shortly after securing a bid to host the 2022 World Cup gave the green light to ExxonMobil to start development of the offshore Barzan gas field project, valued at US$8.6 billion. The major sporting event along with the peninsular state's fast-growing economy will surely call for expansion of its gas-fuelled power generating infrastructure.

To compile this formidable list of oil industry kings, we referred to the much respected ranking metrics of the latest Forbes' annual Global 2000 list of the world's leading companies from 2010 according to their asset values.

As Chinese oil major Sinopec seems to have fallen just outside the ranks of our top 10 compared to the last survey which saw it at tenth place, it only seems fitting that we give the company an honourable mention despite the increasing gap (in terms of asset value) between it and current tenth placeholder ConocoPhillips.[13]

Drawing up a list of the world's top countries that produce the most oil has become a traditional task. It is difficult, however, to predict which ones will sustain their current ranking over the coming years. There is a strong possibility that oil producing powerhouses such as Saudi Arabia, Russia and Iran will be dominating the conventional energy market for years to come.

However, as Iraq slowly rises from the ashes of a post-conflict and post-sanction era, it is increasingly becoming a strong rival. According to the recently published BP 2030 energy outlook, Iraq and Saudi Arabia are set to lead the increase in OPEC's production capacity by 2030. Thus, the importance of OPEC is expected to grow and OPEC's share of global production to increase from 40% in 2010 to 46% in 2030.[14]

The world's top 10 biggest oil reserves are located in Saudi Arabia, Venezuela, Iran, Iraq, Kuwait, the United Arab Emirates, Russia, Libya, Kazakhstan, and Nigeria (Hussain, 2011). In the future, energy – and especially crude oil – is going to flow from one of ten fields. These ten fields are Ghawar (Saudi Arabia),

13 Hussain, Erman. 2011. World's top 10 biggest oil reserves. Arabian oil and gas. February 16th, 2011.

14 BP. 2011. BP Energy Outlook 2030. p. 37.

West Qurna (Iraq), Majnoon (Iraq), Rumaila (Iraq), Khuzestan (Iran), Kashagan (Kazakhstan), Khurais (Saudi Arabia), Tupi (Brazil) Carabobo (Venezuela), and North Slope (Alaska).[15]

2.3 Oil Price History

Crude oil prices behave much as any other commodity with wide price swings in times of shortage or oversupply. The crude oil price cycle may extend over several years, responding to changes in world energy demand.

The U.S. petroleum industry's price was heavily regulated through production or price controls throughout much of the twentieth century. In the post-World War II era, world oil prices at the wellhead have averaged $32.71 per barrel, adjusted for inflation to 2009 dollars. Until the beginning of 2000 (March), the $22-$28 price band for the OPEC basket of crude was adopted, and oil prices only exceeded about $24.00 per barrel in response to war or conflict in the Middle East. With limited spare production capacity, OPEC abandoned its price band in 2005 and was powerless to stem the surge in oil prices.[16]

The very long-term view is much the same. Crude oil prices plummeted below 50 cents/barrel (about $12 in 2009 dollars) by 1861. Since 1861, the world price of crude oil, adjusted for inflation, has averaged $28.63 per barrel in 2009 dollars. Figure 6 depicts the historical development of the oil price.

Fig. 6: World oil prices, 1861-2009

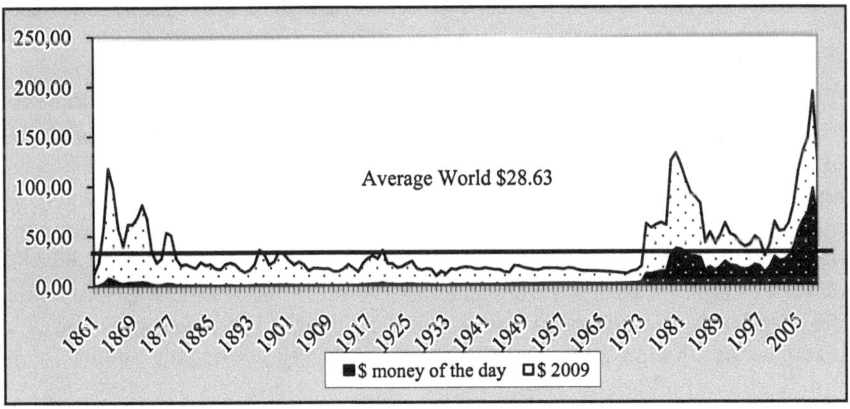

Source: BP. British Petroleum. Statistical Review of World Energy 2010. p. 16.

15 Helman, Christopher. 2010. The World's Biggest Oil Reserves. Forbes.

16 OPEC, Monthly Oil Market Report. 2011.

Since 1947, U.S. oil prices at the wellhead have averaged $26.64 per barrel adjusted for inflation to 2009 dollars. In the absence of price controls, the U.S. price would have tracked the world price, and averaged $28.68. Over the same post-war period, the median for the domestic and the adjusted world price of crude oil was $19.60 in 2009 prices. That means that oil prices exceeded $19.60 per barrel for only fifty percent of the time from 1947 to 2008 (Figure 7).

The results are totally different after 1970, where U.S. crude oil prices averaged $32.36 per barrel and the more relevant world oil price averaged $35.50 per barrel. The median oil price for the same period is $30.04 per barrel.[17]

Fig. 7: Crude Oil Prices 1947- 2009 (Crude Oil Prices in 2009 Dollars)

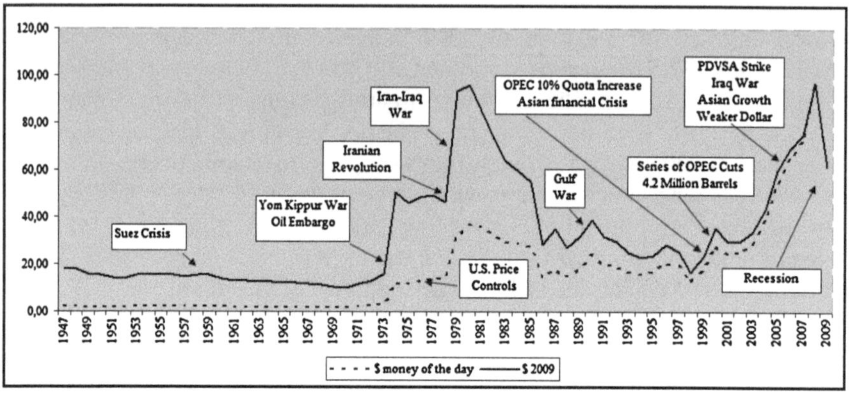

Source: BP. British Petroleum. Statistical Review of World Energy 2010. p. 16.

If oil prices revert to the mean, this period is likely to be the most appropriate for today's analyst. It follows the peak in U.S. oil production eliminating the effects of the Texas Railroad Commission and is a period when the Seven Sisters were no longer able to dominate oil production and prices. It is an era of far more influence for OPEC oil producers than they had in the past. As we will see in the details below, influence over oil prices is not equivalent to control. At least since the oil crises of the 1970s and the dramatic price drop in the mid-1980s, forecasting the oil price development has been a major challenge. The recent turbulences in oil prices in 2008 posed new challenges to analysing the impact parameters on the volatility of oil prices.

The coincidence of oil dates and recessions after 1972 are shown Table 5.

17 WTRG Economics. 2011. Oil Price History and Analysis (Updating).

Table 5: Factors affecting oil prices after 1972

Business cycle peak	Events associated with subsequent major oil price increase
November 1973	Yom Kippur War and oil embargo October 1973-early 1974
January 1980	Iranian revolution October 1978-February 1979
July 1981	Outbreak of Iran-Iraq war September 1980
July 1990	Invasion of Kuwait and Gulf war August 1990
October 1997	Asian financial crisis July 1997
December 1997	OPEC 10% quota increase
March 2001	OPEC Meeting March 1999
December 2003	Petroleos de Venezuela. S. A. strike and Iraq war March 2003
September 2004	Mexico Gulf, Hurricane Ivan September 2004
August 2005	Hurricane Katrina, Hurricane Rita August 2005- September 2005
In the 2000s years	Rapid growth in Asian economies and their petroleum consumption and weaker dollar
In general	Level of petroleum inventories in the U.S. Increasing petroleum demand OPEC quotas

Source: EIA. U.S. Energy Information Administration. 2011. Country Analysis Briefs; WTRG Economics. 2011. Oil Price History and Analysis (Updating).

Factors contributing to the current level of crude oil prices after 2000 include the continued rapid growth in Asian economies and their petroleum consumption, a weak dollar, 2005 hurricanes and U.S. refinery problems, level of petroleum inventories in the U.S. and increasing petroleum demand, and OPEC policies.

3.3 Oil Cartels

An oil cartel is a cartel of companies or nations formed to control the production and distribution of oil. One oil cartel is the Organization of Petroleum-Exporting Countries, known as OPEC. Other cartels are non-Arabian oil companies, informally known as "The Seven Sisters" and "New Seven Sisters".

3.3.1 OPEC

OPEC is an oil cartel with currently 12 member countries in total. The group was founded on September 14, 1960, for the purpose of setting oil prices by controlling oil production. They were originally thought to be primarily Arabian in ownership, however, it is actually an international organization, which includes Americans.[18]

3.3.2 The Seven Sisters

Another cartel is The Seven Sisters, composed of non-Arabian oil companies who control what is shipped to the United States, and how much is refined into gas and heating oil. Originally, it was made up of the Rockefeller-controlled Exxon (previously known as Standard Oil of New Jersey, or Esso), Mobil (Standard Oil of New York, which merged with Vacuum Oil), and Chevron (Socal, or Standard Oil of California); the Mellon's Gulf Oil; Shell, Texaco, and BP. They controlled 90% of all crude oil exports to world markets by controlling every important pipeline in the world, such as the 753-mile TransArabian Pipeline, from Qaisuma in Saudi Arabia to the Mediterranean Sea, which was owned by Exxon, Chevron, Texaco, and Mobil. Exxon owned the 100-mile Interprovincial Pipeline in Canada, and also the 143-mile pipeline in Venezuela. The 799-mile Alaskan Pipeline was owned by BP and Exxon. By controlling these, and other vital arteries, they can restrict the flow of oil, limiting supplies to refineries.[19]

Mergers and takeovers, particularly from 1998 to 2001, strengthened the grip of these companies so that the 'Seven' became the 'Four'. The first of the Sisters to fall was Gulf, taken over by Chevron in 1984. Exxon and Mobil merged in 1998 to become the largest oil company, but in 2010 it become four companies. BP took over three other important oil companies, Amoco, Arco and Burmah-Castrol in 1998-2000 to take second place, while Chevron and Texaco merged in 2001 to become the number eight company. A new European grouping created from the merger of two French oil companies, Total and Elf, and the Belgian company, Fina, created a sixth world power. Two large US companies, Conoco and Phillips merged in 2002 to form the largest company, but it fell to tenth place in 2010. There is then a gap in size down to the Italian company, ENI and the Spanish company, Repsol.[20]

---

18 OPEC. 2011. About Us.

19 Rivera, David Allen. 2004. Final Warning: A History of the New World Order. Illuminism and the master plan for world domination; Bamberg, James. 1994.

20 Thomas, Steve. 2003. "Viewpoint The Seven Brothers", Energy Policy 31 (2003).

3.3.3 New Seven Sisters

Finally, the Financial Times has used the label the "New Seven Sisters" to describe a group of what it argues are the most influential national oil and gas companies based in countries outside of the OECD. These companies consist of Saudi Aramco (Saudi Arabia), Gazprom (Russia), China National Petroleum Corporation (China), National Iranian Oil Company (Iran), Petroleos de Venezuela. S.A (Venezuela), Petrobras (Brazil), and Petronas (Malaysia).

Overwhelmingly state-owned, these companies control almost one-third of the world's oil and gas production and more than one-third of its total oil and gas reserves. In contrast, the old seven sisters produce about 10% of the world's oil and gas and hold just 3% of reserves. Even so, their integrated status pushes their revenues notably higher than those of the newcomers.[21]

4. OPEC Oil Reserves, Production

OPEC was founded in Baghdad, Iraq, with the signing of an agreement in September 1960 by five countries, namely the Islamic Republic of Iran, Iraq, Kuwait, Saudi Arabia and Venezuela. They were to become the Founder Members of the Organization. These countries were later joined by Qatar (1961), Indonesia (1962), Libya (1962), the United Arab Emirates (1967), Algeria (1969), Nigeria (1971), Ecuador (1973), Gabon (1975) and Angola (2007). From December 1992 until October 2007, Ecuador suspended its membership; and Gabon terminated its membership in 1995. Indonesia suspended its membership effective January 2009. Currently, the Organization has a total of 12 member countries.

The purpose of OPEC, as with any cartel, is to raise oil prices by controlling oil production. The oil industry has been plagued by production booms and falling prices ever since Colonel Drakes' discovery of oil at Titusville, Pennsylvania in 1859. Just as the major oil companies colluded from the 1920s to the 1960s to prevent prices from falling, members of OPEC meet on a regular basis to set production levels in the hope of maintaining prices. The essential nature of oil coupled with its limited number of suppliers makes it the ideal product for cartelization.[22]

21 Hoyos, Carola. 2007. "The new Seven Sisters: oil and gas giants dwarf western rivals". The Financial Times, March 11, 2007.

22 OPEC. 2011. About Us.

4.1 Brief History of OPEC

OPEC is an international cartel of oil-producing states that has attempted – with varying degrees of success – to manipulate world oil prices over the past fifty years.

OPEC is a permanent, intergovernmental Organization, created at the Baghdad Conference on 10-14 September 1960, by Iran, Iraq, Kuwait, Saudi Arabia and Venezuela. These countries were later joined by Qatar (1961), Indonesia (1962, suspended its membership from January 2009), Libya (1962), United Arab Emirates (1967), Algeria (1969), Nigeria (1971), Ecuador (1973, suspended its membership from December 1992-October 2007), Angola (2007), and Gabon (1975-1994). Currently, the organization has a total of 12 member countries. OPEC had its headquarters in Geneva, Switzerland, in the first five years of its existence, but moved to Vienna, Austria, on 1 September 1965.

These countries were a group of major oil producing countries who wished to coordinate national petroleum policies and regulate a more united front in dealings with the multinational oil companies who were licensed to produce and export petroleum from their lands. In those years, the combined membership of OPEC accounted for over half of worldwide crude oil production.

OPEC's purposes is to coordinate and unify petroleum policies among member countries, in order to secure fair and stable prices for petroleum producers; an efficient, economic and regular supply of petroleum to consuming nations; and a fair return on capital to those investing in the industry.

The 1960s: OPEC's formation by five oil-producing developing countries in 1960 occurred at a time of transition in the international economic and political landscape, with extensive decolonization and the birth of many new independent states in the developing world. In the 1960s, the international oil market was dominated by the multinational "Seven Sisters" companies and was largely separate from that of the former Soviet Union and other centrally planned economies. OPEC established its Secretariat, first in Geneva and then, in 1965, in Vienna. It adopted a "Declaratory Statement of Petroleum Policy in Member Countries" in 1968.[23]

The 1970s: OPEC rose to international prominence during this decade, as its member countries took control of their domestic petroleum industries and acquired a major say in the pricing of crude oil on world markets. On two occasions, oil prices rose steeply in a volatile market, triggered by the Arab oil embargo in 1973 and the outbreak of the Iranian Revolution in 1979. Membership grew to 13 by 1975.

23 OPEC. 2011. About Us.

The 1980s: After reaching record levels early in the decade, prices began to weaken, before crashing in 1986, responding to a big oil glut and consumer shift away from this crude oil. OPEC's share of the smaller oil market fell heavily (28.5% in 1985) and its total petroleum revenue dropped. But in the final part of the decade, the price of crude oil increased, and OPEC's share of newly growing world output began to recover. This was supported by OPEC introducing a group production ceiling divided among member countries and a reference basket for pricing.

After 1980, oil prices began a six-year decline that culminated with a 46 percent price drop in 1986. This was due to reduced demand and over-production that produced a glut on the world market. Around this period, Iraq also increased its oil production to help pay for the Iran-Iraq War. Overall OPEC lost its unity and thus its net oil export revenues fell in the 1980s.

The 1990s: Crude oil prices moved less dramatically than in the 1970s and 1980s. In the decade, excessive volatility and general price weakness dominated, and the South-East Asian economic downturn and mild Northern Hemisphere winter of 1998-99 saw prices back at 1986 levels. However, a solid recovery followed in a more integrated oil market, which was adjusting to the post-Soviet world, greater regionalism, globalization, the communications revolution and other high-tech trends. In the 1990s, Gabon left OPEC, while Ecuador suspended its membership.

The 2000s: An innovative OPEC oil price band mechanism helped strengthen and stabilize crude prices in the early years of the decade. But a combination of market forces, speculation and other factors transformed the situation in 2004, pushing up prices and increasing volatility in a well-supplied crude market. Oil was used increasingly as an asset class. Prices soared to record levels in mid-2008, before collapsing in the emerging global financial turmoil and economic recession. OPEC became prominent in supporting the oil sector, as part of global efforts to address the economic crisis. OPEC's second and third summits in Caracas and Riyadh in 2000 and 2007 established stable energy markets, sustainable development and the environment as three guiding themes, and it adopted a comprehensive long-term strategy in 2005.[24]

4.2 OPEC Oil Production and Quota

OPEC's ability to hold the price of oil above the competitive level is dependent upon barriers to entry, which in this case hinge upon OPEC's dominant ownership and control of low-cost oil reserves. By chance of nature, some 77% of the

24 OPEC. 2011. About Us.

world's proven reserves of crude oil are located in OPEC nations.[25] Proven reserves constitute that portion of the ultimate resource which has been discovered and is commercially producible. Additional reserves can be developed through exploration, discovery, and development of new fields, but this process has become increasingly difficult and expensive. Thus, while production of crude oil from non-OPEC sources does expand in response to the higher prices that result when cartel members restrict output, the scope for this is limited and will remain so. Moreover, OPEC's efforts to manipulate the price of oil are protected from anti-trust enforcement and legal intervention by the sovereign rights of its members.[26]

Whatever one's opinions on those matters are, there is no question that OPEC members have restricted production in ways that are unrelated to the inherent scarcity of crude oil. Although OPEC's proven oil reserves were steadily rising between 1973 and 1985, production was cut by nearly half during that twelve year interval, falling from about 30 million barrels per day (mbpd) in 1973 to an all-time low of 15.4 mbpd in 1985. Today, OPEC continues to hold production below the 1973 level, although the proven oil reserves of OPEC members have doubled in volume since then.

OPEC accounts for a significant portion of world crude oil production (Figure 8). OPEC's share of world oil production increased in some years, and declined in others (Figure 14).

OPEC did not manage to increase prices in the 1960s, but it did manage to hold the line on taxes. When oil production in the United States peaked in 1970 and markets tightened, OPEC was able to raise prices during the Yom Kippur war in 1973[27]. Prices rose again in 1979 after the Iranian revolution. These rapid price increases are seen in Figure 11. They caused a reduction in oil consumption and an increase in production outside of OPEC. OPEC's production fell by about a third between 1973 and 1985, and OPEC's share of world oil markets fell from more than 53% to about 28%. OPEC's share has rebounded somewhat since 1985.

---

25 BP Energy Outlook 2030.

26 OPEC. 2011. About Us.

27 The persistence of the Arab-Israeli conflict finally triggered a response that transformed OPEC into a formidable political force. After the Six Day War (June War or Third Arab-Israeli War) of 1967, the Arab members of OPEC formed a separate, overlapping group, for the purpose of centering policy and exerting pressure on the West over its support of Israel. Egypt and Syria, though not major oil-exporting countries, joined the latter grouping to help articulate its objectives. Later, the Yom Kippur War of 1973 galvanized Arab opinion. Furious at the emergency re-supply effort that had enabled Israel to withstand Egyptian and Syrian forces, the Arab world imposed the 1973 oil embargo against the United States and Western Europe, while non-Arab OPEC members did not (Basil, 2011).

Fig. 8: World and OPEC Oil Production

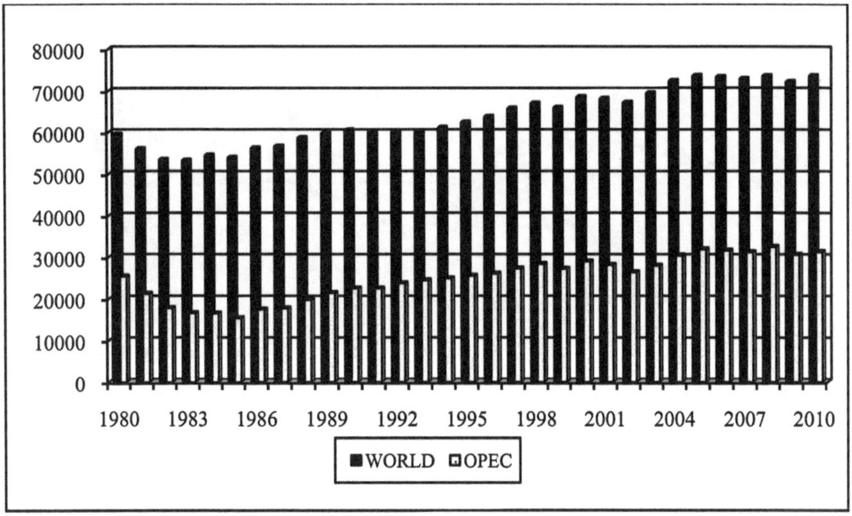

Source: EIA. U.S. Energy Information Administration. 2011. Country Analysis Briefs.

Saudi Arabia, in particular, took a large hit, with production falling from a peak of more than 9 million barrels a day in the late 1970s and early 1980s to less than 3.5 million barrels a day in 1985. As a result, in late 1985, it began using netback pricing[28], and in 1986, it raised production to more than 4.5 million barrels a day. As a result, world prices plummeted from $27.56 to $14.43 per barrel in 1986.

Oil prices remained at lower levels with a slight downward trend, expect for the Gulf War blip in 1990-1991. At lower prices, demand increased and non-OPEC production decreased. OPEC sales climbed, as did its share of world oil production (Figure 9).

28 Netback price is a price a refinery receives for the sale of petroleum products after deducting the transportation or affiliated costs in shipping the product from its point of origin (i.e., pipeline tariffs, waterborne freight, storage fees, line loss, cost of capital, etc.) (OPIS, 2011). The netback method for pricing crude oil, although used in the past by integrated companies in valuing oil transferred between subsidiaries, has become the subject of much interest and controversy after its adoption by Saudi Arabia during the second half of 1985 (Mabro, 1987).

Fig. 9: OPEC's share of world oil production

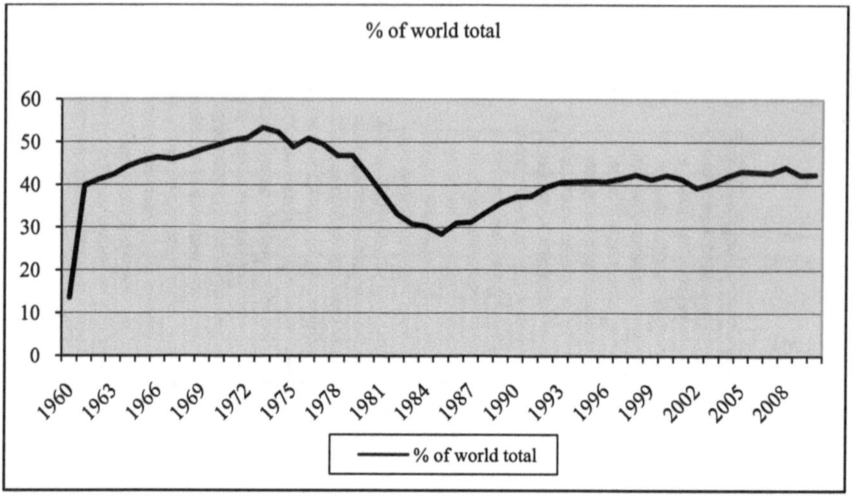

Source: EIA. U.S. Energy Information Administration. 2011. Country Analysis Briefs.

Oil demand growth remained strong, and discipline within OPEC remained reasonably good until after 1996. Continued growth in production in 1997 and early 1998 was largely the result of quota violation. Venezuela overproduced the most (Figure 10).

Throughout the 1990s, OPEC exceeded its production ceiling by more than 1 million barrels a day. This, along with the decrease for oil demand in the world due, for example, to the Asian financial crisis and the 10% OPEC 10% quota increase (Figure 7), again caused prices to fall below $12.7 per barrel during 1998. In early 1999, the tide started to turn, as a new President was elected in Venezuela. Under his leadership, OPEC pulled together to reduce production. Gradually, the Asian economy picked up. Low oil prices had taken their toll on higher-cost producers. The market tightened, pushing oil prices to more than $20 per barrel during 2000 with an average of around $30. The U.S. economy was weaker in 2001 than in 2000, and oil prices were listed at more than $25 for most of the first part of the year. They fell below $25 after the U.S. World Trade Center in New York was attacked on September 11, and third-quarter U.S. gross domestic product fell. Prices averaged around $24.4 for the year 2001.[29]

29 Dahl, Carol Ann. 2004. International Energy Markets, Understanding Pricing, Policies, and Profits. Tulsa, Oklahoma: PennWell Press. p. 151.

Fig. 10: OPEC Average Production and Quotas, 1982-2009

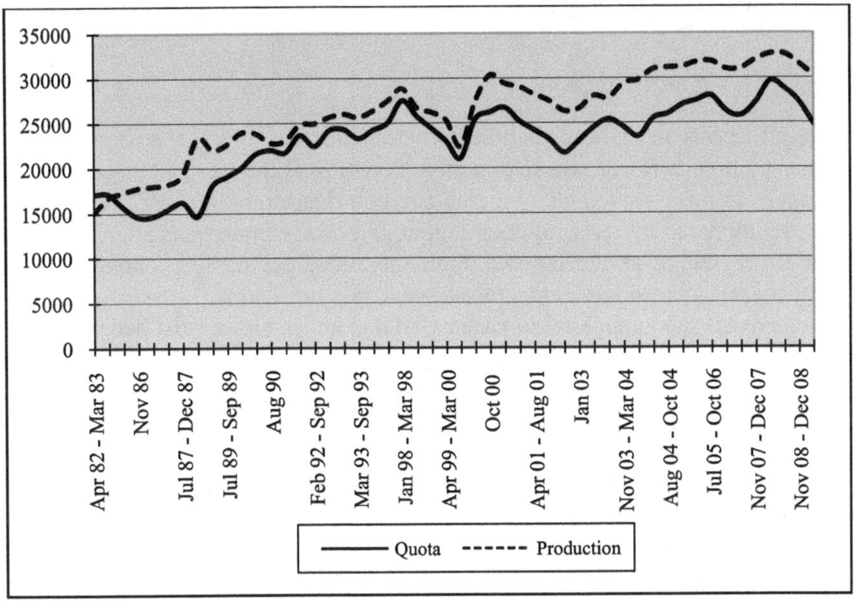

Source: Dahl, Carol Ann. 2004. International Energy Markets, Understanding Pricing, Policies, and Profits. Tulsa, Oklahoma: PennWell Press. p. 149-150: OPEC, 2011. About Us.

After the 2001 attacks on the United States, the following invasion of Afghanistan, and the 2003 invasion of Iraq and subsequent occupation prompted a surge in oil prices to levels far higher than those targeted by OPEC during the preceding period. Indonesia withdrew from OPEC to protect its oil supply interests. On 19 November 2007, global oil prices reacted strongly as OPEC members spoke openly about potentially converting their cash reserves into Euro and out of US dollars.[30]

Figure 10 shows this event in real U.S. oil prices. Initially, turbulent prices stabilized when Rockefeller gained monopoly power, and they remained stable when the Seven Sisters and Texas Railroad Commission reigned, from 1930 to 1970. There was turbulence as OPEC took over until 1986, when market pressures moved OPEC to a quota system and prices reverted to historic levels. Currently, OPEC is riding high, but weathering the Asian crisis may not be enough.

30 Basil, Ajit. 2011. OPEC & its influence on Price of Oil. Energy Sector Structure Policies & Regulations. Submitted to Prof. Anupam Sircar toward End-Term Assessment. 14-Jan-11.

Global warming concerns and ensuing regulation may be just around the corner to provide further challenges.[31]

4.3 OPEC's Future Prospects

Several aspects must be considered when evaluating OPEC's past success or contemplating its future course of action. Foremost is the fact that any system of output restraint is vulnerable to the classic free rider problem. OPEC as a whole may be improved by reducing total output, but each member has an incentive to produce beyond its assigned quota. From the individual member's point of view, marginal revenue from incremental sales exceeds the marginal cost of extraction, which creates the temptation to cheat. Cartel membership is most beneficial to a producer when other members are doing the hard work. But without a system to detect and punish cheating, the cartel is prevented by a prisoner's dilemma in which the dominant strategy for most, if not all, members is to ignore their assigned production quotas.

In fact, OPEC lacks an effective means to monitor, detect, and punish members who exceed their quotas. A monthly chart of OPEC's combined crude oil production level relative to the agreed ceiling indicates the discrete and continuous nature of this problem (Figure 15). As seen in the figure, production does sometimes comply with the quota. Since the beginning of the quota system, total OPEC production of crude oil has exceeded the ceiling by 4% on average, but on numerous occasions the excess has run to 15% or more. For the most part, compliance has been achieved only during episodes when the production ceiling itself pushed the limits of each member's available production capacity.

A second factor that confounds OPEC's attempt to manage the market price is the lack of timely and accurate information concerning changes in the level of demand for oil. Several forecasts of demand and supply are available at any given time, but the precision of these forecasts is low and surprises are frequent. Even if perfect information about future market conditions were available, there is no assurance that the interests of individual OPEC members could be easily aligned around a single "correct" price or production target.[32]

Therefore, any given set of quotas determines both the overall profit of OPEC and the individual revenues that accrue to each member. If the members were more homogeneous demographically and economically, the problem of misaligned interests would be less severe. As it happens, however, large volumes

---

31 Dahl, Carol Ann. 2004. International Energy Markets, Understanding Pricing, Policies, and Profits. Tulsa, Oklahoma: PennWell Press. p. 151.

32 Smith, James L. 2007. The Role of OPEC in the World Oil Market. Submission to the joint Economic Committee, U.S. Congress, May 23, 2007. p. 7.

of low-cost reserves are concentrated in certain countries with small populations and relatively high incomes like Kuwait, Saudi Arabia, and the United Arab Emirates, while smaller volumes of higher-cost reserves are found in populous and relatively poor countries like Nigeria, Indonesia, and Venezuela. Table 6 sets forth some of the more distinct differences among the members of OPEC. The potential for conflicting interests involves not only the question of which members deserve larger quotas, but what is the preferred market price level for OPEC oil.

The final factor affecting OPEC's future is the role played by unforeseen incidents and geopolitical issues. The obvious historical impact of OPEC on oil prices has originated not only from the result of wilful setups made by purposeful trusts, but also from the sparks of clashing agendas of related nations.[33] During the past fifty years, most of the idle oil production capacity of OPEC member nations has been involuntarily – taken out of production due to military conflicts. Thus, most of the hard work to be done by OPEC to balance oil prices has been arranged fortuitously by the parties of armed conflict by restricting the output of production. Consequently, the actual power and cohesive force of OPEC has not yet been actually proved.

Table 6: Differences among OPEC Members, 2009

Member	since	GDP $ per capita	Value of oil export $ per capita	Proved oil reserves bbl per capita	Crude oil production bbl per capita	Reserves to production ratio years
Algeria	1969	3,751	864	345	34	10
Angola	2007	3,717	2,098	514	94	5
Ecuador	1973	3,669	497	465	33	14
Iran*	1960	4,460	750	1,849	48	39
Iraq*	1960	4,363	1,340	3,682	75	49
Kuwait*	1960	30,240	13,382	29,167	650	45
Libya	1962	10,479	4,895	7,242	230	31
Nigeria	1971	1,064	170	238	12	20
Qatar	1961	51,201	16,366	15,756	447	35
Saudi Arabia*	1960	14,552	6,197	10,417	322	32
U.A.E.	1967	49,782	12,697	21,169	485	44
Venezuela*	1960	11,863	1,910	7,441	101	73
Total OPEC		5,208	1,441	2,666	72	37

* *Founding member of OPEC*
Source. OPEC. 2009. Annual Statistical Bulletin 2009. p. 11.

33 Smith, James L. 2007. The Role of OPEC in the World Oil Market. Submission to the joint Economic Committee, U.S. Congress, May 23, 2007. p. 8.

The daily value of the crude oil that is produced and sold in the world markets exceeds $1 billion. Even a relatively small deviation in the unit oil price means a huge cumulative wealth transfer between the parties such as consumers and producers. Moreover, the subversive impact of sudden fluctuations and the unbalanced nature of the oil price jeopardize the goal of sustainable global economic growth. In this volatile environment in which consumers, investors, and government officials continue to challenge these problems, it can be said that OPEC has a significant impact on the world economy through its role in determining the price of oil if it acts wisely.[34]

4.4 OPEC's Position in Trading Spot & Oil Futures

OPEC member countries account for approximately 42% of the world's crude oil production. OPEC's crude oil exports represent 58% of the crude oil traded internationally. Therefore, OPEC has a significant influence on world oil markets.[35]

Another fact that significantly strengthens OPEC's position is that the OPEC members are the only nations with significant spare oil production capacity among the entire world's oil producing countries. Since they have approximately 80% of the world's oil reserves, OPEC nations can expand oil production when demand increases. They respond to market fundamentals and forecast developments by coordinating their petroleum policies. If demand grows or some producers are producing less oil, OPEC can increase its oil production in order to prevent a sudden rise in prices. [36]

But increasing oil production is not an easy procedure. Before expanding their output, the member countries need to be sure that the oil industry will continue to be profitable after this expansion. Oil producers invest billions of dollars in explorations and infrastructures. A new field can take 3 to 10 years to locate and develop. Sometimes when it is not used or used at less than optimum capacity for a long period of time, the plant may suffer from irreparable damage. Shutting down a plant in the middle of the ocean due to lack of oil demand is also difficult and incurs a giant loss. If oil producers do not invest enough money and/or do not invest it far enough in advance, then the world could face a shortage of oil supplies in future, which is also influencing the oil price.

---

34 Smith, James L. 2007. The Role of OPEC in the World Oil Market. Submission to the joint Economic Committee, U.S. Congress, May 23, 2007. p. 9.

35 EIA. U.S. Energy Information Administration. 2010. International Energy Outlook 2010; BP. Statistical Review of World Energy 2010.

36 Chanda, Anita. What is OPEC and how it influences the Oil Price. Indian Blogger. August 23, 2010.

OPEC set crude oil prices from the early 1970s to mid-1980s, but does not do so today. The price of crude oil is set by movements on three major international petroleum exchanges – the New York Mercantile Exchange (NYMEX), the International Petroleum Exchange in London (IPE) and the Singapore International Monetary Exchange (SIMEX). The International Energy Agency (IEA) in Paris and US Energy Information Administration (EIA) are also important players. Spot and future trading in these markets affects the oil prices.[37]

Most of the oil futures of different blends produced by OPEC are traded on The Dubai Mercantile Exchange (DME). The DME is the premier energy-focused commodities exchange in the East of Suez and is home to the world's third crude benchmark. The DME was launched in June 2007 with the goal of bringing fair and transparent price setting and efficient risk management to the East of Suez (Asia Pacific and Middle East), the world's fastest growing commodities market and already the largest crude oil supply/demand corridor in the world.

The DME lists the Oman Crude Oil Futures Contract (DME Oman) as its flagship contract, providing the fairest and most transparent crude oil benchmark in the East of Suez. Today, DME Oman is the explicit and sole benchmark for Oman and Dubai crude oil Official Selling Prices – the historically established markers for Middle East crude oil exports to Asia.

The DME is regulated by the Dubai Financial Services Authority (DFSA) and all trades executed on the exchange are cleared through and guaranteed by the New York Mercantile Exchange (NYMEX), a member of the CME Group, which is regulated by the U.S. Commodity Futures Trading Commission (CFTC) and is a Recognized Body by the DFSA. The DME is recognized by over 20 international regulatory bodies across Europe and Asia, further cementing its position as truly global and well-regulated marketplace.[38]

Even though today OPEC is not entirely responsible for oil prices, OPEC member countries do voluntarily restrain their crude oil production in order to stabilize the oil market. Since the oil industry is the sole source of revenue for most of the OPEC nations, they try to stabilize the oil market to maintain profits and the economic growth of the member countries. In 2008-09 global oil demand declined in two successive years for the first time since 1981 due to global recession. Demand fell by 1.8 million barrels per day and the price of a barrel of crude lost almost 100$ in less than 6 months from mid-2008, resulting in unused production capacity.[39]

---

37 Chanda, A. What is OPEC and how it influences the Oil Price. Indian Blogger. August 23, 2010.

38 DME Dubai Mercantile Exchange. About DME. 2011.

39 Chanda, Anita. What is OPEC and how it influences the Oil Price. Indian Blogger. August 23, 2010; Basil, Ajith. 2011. OPEC & its influence on Price of Oil. Energy Sector Structure Policies & Regulations. Submitted to Prof. Anupam Sircar toward End-Term Assessment. 14-Jan-11.

4.5 Impact of Kyoto Protocol on OPEC

Exploring alternate sources of energy has been gaining popularity since environmental concerns began significantly occupying mankind. The possibility of a drop in the oil sector caused by these non-conventional sources has started the green paradox dispute.[40] The assumption of a gradual reduction in global oil consumption causing a decline in future oil prices is posing a threat for oil producing nations. The dynamic behaviour of the interrelated variables shows that if this assumed drop in oil prices affects the economic stability of oil producers, they may increase production to balance their economic indicators and sustain their oil-based revenue, a manoeuvre that may result in increased oil consumption due to low prices and high supply in the market. Consequently, this change in the market may accelerate global warming in the short run. The reaction of oil producing countries to the sustainable energy sources revolution may be more disastrous, as OPEC has stated that the nations that depend strongly on oil exports need to be compensated for the negative effects created by the consequences of the KYOTO Protocol.[41]

Conclusion

OPEC has played an important role in international oil markets for five decades. It seems that this important role will continue into the future. We have examined this role more closely and addressed some opinions on OPEC's revising its vision for the new millennium.

Because OPEC member nations account for the majority of crude oil production and international trade, OPEC has a significant influence on world oil markets. OPEC member countries produced about 42% of the world's crude oil in 2010. OPEC's crude oil exports represent a significant 58% of the crude oil traded internationally.

Different scenarios agree that energy demand will continue to increase in the future. According to OPEC's World Oil Outlook reference case, it will increase about 50% by 2030. Renewable energy sources will grow fast, but from a low base. Realistically, fossil fuels will continue to fulfil most of the world's energy needs, contributing more than 80% to the global energy mix over this period.

---

40 Sinn, Hans-Werner. The Green Paradox. Global Warming: The Neglected Supply Side. 25 November 2008. DG Ecofin.

41 Basil, Ajith. 2011. OPEC & its influence on Price of Oil. Energy Sector Structure Policies & Regulations. Submitted to Prof. Anupam Sircar toward End-Term Assessment. 14-Jan-11.

Among all other fuel types, oil will continue to play the leading role, although its share will be slightly lower in 2030.

Meeting the demand of oil consumers, working against global climate change and optimizing oil capacity investment plans to maintain global economic balances are likely to be the predominant issues for consideration by all of the nations and international organizations in the process of regulating crude oil production and prices.

Resources

Bamberg, J. (1994). The History of the British Petroleum Company. Volume 2 The Anglo-Iranian Years, 1928-1954: The British Petroleum Company plc. http://books.google.com.tr/books?id=PduotC73nh0C&pg=PA107&lpg=PA107&dq=Achnacarry+Agreement,1928&source=bl&ots=arSIJGOp1c&sig=-26hLwz9mUSqehA_BqrEXIqvXRY&hl=tr&ei=OrG_TamPB87ysgap0t3DBQ&sa=X&oi=book_result&ct=result&resnum=5&ved=0CEQQ6AEwBA#v= one page&q=Achnacarry%20Agreement%2C1928&f=false. (28.04.2011).

Basil, A. (2011). OPEC & its influence on Price of Oil. Energy Sector Structure Policies & Regulations. Submitted to Prof. Anupam Sircar toward End-Term Assessment. 14-Jan-11.www.iemr.in/.../OPEC%20&%20its%20Influence%20on%20the%20Price%20of%20Oil_ (09.05.2011).

BP. (2010) British Petroleum. Statistical Review of World Energy 2010. http://www.bp.com/productlanding.do?categoryId=6929&contentId=7044622 (22.04.2011).

BP. (2011). BP Energy Outlook 2030. http://www.bp.com/sectiongenericarticle.do?categoryId=9035979&contentId=7066648 (30.04.2011).

Chanda, A. (2010). What is OPEC and how it influences the Oil Price. Indian Blogger. August 23, 2010, http://indianblogger.com/world-oil-price-opec/ (09.05.2011).

Chon, G. (2009). Iraq Awards West Qurna-1 Oil Field to Exxon, Shell. The Wall Street Journal. November 5, 2009. http://online.wsj.com/article/SB125741092983330293.html (12.05.2011).

Dahl, C.A. (2004). International Energy Markets, Understanding Pricing, Policies, and Profits. Tulsa, Oklahoma: PennWell Press.

DME Dubai Mercantile Exchange. (2011). About DME. http://www.dubaimerc.com/about-us/about_dme.aspx (09.05.2011).

Edwards, J.D. (2001). Twenty-first-Century Energy: Decline of Fossil Fuel, Increase of Renewable Nonpolluting Energy Sources. In M. V. Downey, J.C. Threet and W. A. Morgan (Edited) Petroleum Provinces of the Twenty-first

Century (AAPG Memoir 74): 21-34. http://books.google.com.tr/books?id=i9jts14tq1gC&pg=PA21&lpg=PA21&dq=%22Increasing+World+Energy+Demand%22&source=bl&ots=S_x5X-synm&sig=mtogZNJarXEoCkiSdDaX23ttdks&hl=tr&ei=eQ-xTYrnNIbKswaG9ajdCw&sa=X&oi=book_result&ct=re sult&resnum=7&ved=0CE8Q6AEwBg#v=onepage&q=%22Increasing%20World%20Energy%20Demand%22&f=false (22.04.20111).

EIA. (2010). U.S. Energy Information Administration. 2010. International Energy Outlook 2010. http://www.eia.doe.gov/oiaf/ieo/ieorefcase.html (12.03.2011).

EIA. (2011). U.S. Energy Information Administration. Country Analysis Briefs. http://www.eia.doe.gov/countries/ (12.03.2011).

EnerBridge. (2011). Statistics of the Oil and Gas World. EnerBridge, Smart Solutions for Oil & Gas Professionals. http://www.enerbridge.biz/statistics.html#BiggestCompaniesbyReserves (15.04.2011).

Gelsi, S. (2010). Venezuela rolls out welcome for foreign oil companies. Carabobo region to draw bidders from Western oil giants. The Wall Street Journal. MarketWatch. January, 24. http://www.marketwatch.com/story/venezuelas-carabobo-region-drawing-big-oil-2010-01-24 (12.05.2011).

Grant, W. E. and Dvorak, K. (2011). The American 1890s: a Chronology. John D. And Standard Oil. http://www.bgsu.edu/departments/acs/1890s/rockefeller/bio2.htm (15.04.2011).

Greg Croft Inc. (2011). The Ghawar Oil Field, Saudi Arabia. http://www.gregcroft.com/ghawar.ivnu (12.05.2011).

Helman, C. (2010). The World's Biggest Oil Reserves. Forbes. http://www.forbes.com/2010/01/21/biggest-oil-fields-business-energy-oil-fields.html. (15.03.2011).

Hoyos, C. (2007). The new Seven Sisters: oil and gas giants dwarf western rivals. The Financial Times, March 11, 2007. http://www.ft.com/cms/s/2/471ae1b8-d001-11db-94cb-000b5df10621.html#axzz1KnpxvUuw (14.03.2011).

Hussain, E. (2011). World's top 10 biggest oil reserves. Arabian oil and gas. February 16th, 2011. http://www.arabianoilandgas.com/article-8344-new-worlds-top-10-biggest-oil-reserves/ (28.04.2011).

IEA. (2009). International Energy Agency. World Energy Outlook 2009. http://www.worldenergyoutlook.org/2009.asp (15.03.2011).

IEA. (2010). International Energy Agency. Key World Energy Statistic 2010. http://www.iea.org/publications/free_new_Desc.asp?PUBS_ID=1199 (15.03.2011).

IMF. (2011). International Monetary Fund. World Economic and Financial Surveys World Economic Outlook (WEO). Tensions from the Two-Speed Recovery: Unemployment, Commodities, and Capital Flows. April 2011. http://www.imf.org/external/pubs/ft/weo/2011/01/index.htm (22.04.2011).

Live Oil Prices. (2011). About Crude Oil. http://www.liveoilprices.co.uk/crude_oil/index.html (10.04.2011).

Lukman, R. (2000). The Role of OPEC in the 21st Century. World energy, 3(1). www.eppo.go.th/inter/opec/RoleOfOPEC.html (18.02.2011).
Mabro, R. (1987). Netback Pricing and the Oil Price Collapse of 1986. Oxford Institute for Energy Studies. WPM 10. January 1987. www.oxfordenergy.org/pdfs/WPM10.pdf(15.03.2011).
OPEC. (1961). Statute. The original text of the organization's statute was approved by the Conference in January 1961 in Caracas – Resolution II.6.
OPEC. (2009). Annual Statistical Bulletin 2009. http://www.opec.org/opec_web/en/publications/202.htm (15.03.2011).
OPEC. (2011). About Us. http://www.opec.org/opec_web/en/. (22.04.2011).
OPIS Pricing. (2011). News & Analysis from the Refinery to the Pump. OPIS Energy Glossary. http://www.opisnet.com/market/glossary.asp#N (12.04.2011).
Petrostrategies. (2011). World's Largest Oil and Gas Companies. http://www.petrostrategies.org/Links/worlds_largest_oil_and_gas_companies.htm (12.03.2011).
PRB. (2010). 2010 World Population Data Sheet: In English. http://www.prb.org/Articles/2010/worldpopulationclock2010.aspx (22.04.2011).
Rivera, D.A. (2004). Final Warning: A History of the New World Order. Illuminism and the master plan for world domination. http://www.coming-judgement.250x.com/index_files/FW_files/fw8.htm (24.04.2011).
Robelius, F. (2005). Giant Oil Fields of the World. Presentation AIM Industrial Contact Day, Monday 23rd May 2005. www.peakoil.net/AIMseminar/UU_AIM_Robelius.pdf. (15.03.2011).
Saudi Aramco. (2011). Energy to the World. http://www.saudiaramco.com/irj/go/km/docs/SaudiAramcoPublic/ExternalFiles/ETTW/SA_Home.swf (02.04.2011).
Sinn, H.W. (2008). The Green Paradox. Global Warming: The Neglected Supply Side. 25 November 2008. DG Ecofin. http://www.cesifo-group.de/portal/page/portal/ifoHome/f-about/f3aboutifo/20ifovideo/_ifovideo?item_link=vortrag-sinn-251108.htm (09.05.2011).
Smith, J.L. (2007). The Role of OPEC in the World Oil Market. Submission to the joint Economic Committee, U.S. Congress, May 23, 2007. www.house.gov/jec/hearings/testimony/110/05-23-07/Smith.pdf (14.03.2011).
Thomas, S. (2003). Viewpoint The Seven Brothers. Energy Policy 31: 393-403. http://fourfact.com/images/uploads/SEVEN_Brothers.pdf. (12.02.2011).
U.S. Census Bureau. (2011). World Population Summary. http://www.census.gov/ipc/www/idb/worldpoptotal.php (22.04.2011)
Wikipedia. (2011). List of Oil Fields. http://en.wikipedia.org/wiki/List_of_oil_fields#cite_note-Oil_Supply_and_Demand-5 (12.03.2011).
WTRG Economics. (2011). Oil Price History and Analysis (Updating). http://www.wtrg.com/prices.htm. (12.03.2011).

Energy Deficit and Dependency of Turkey

Yusuf Bayraktutan, Hakki M. Ay and Nazan Şahbaz

Introduction

Satisfying the unlimited needs of individuals and societies with goods and services requires production factors such as labour, capital, entrepreneurship, natural resources and, most notably, energy. In primitive societies, before the discovery of energy resources, manpower was sufficient for the activities of gathering and hunting. Together with industrialization, the discovery of fire and the power of water, and their use in the production process, increased the importance of energy. In the modern world, complex tools and utilities strengthened the role of energy.

There is an obvious gap between current world supply and demand of energy. As a developing country, the known hydrocarbon resources are clearly insufficient to meet the gradually growing energy demand of Turkey. On the one hand, Turkey is concentrating on developing its domestic resources; on the other hand, it is trying to mobilize possible potential, such as introducing renewable energy resources, increasing energy utilization efficiency, and taking advantage of the country's geographical location.

This study analyses the energy demand and supply of Turkey before discussing the country's dependency on external sources and related issues.

1. Energy Consumption

Despite the intensive search for alternatives, for both technical and economic reasons, the three fossil fuels – coal, oil, and natural gas – continue to dominate. As of the beginning of 2008, oil accounted for 35.6%, coal for 28.6%, natural gas for 23.8%, renewable resources for 6.4%, and nuclear energy for 5.6% of world primary energy consumption. Turkish energy consumption comprises virtually the same weighting of fossil fuels as the rest of the world. In total consumption, oil takes a share of 33%, natural gas 29%, coal 26%, renewable resources 4%, and the others 8%.[1]

1 Necdet Pamir (2008). "Energy Policies of the Republic of Turkey Today", **Turkish Economy,** Ed: Gülen E. ARSLAN, Gazi Y., Ankara. 93-159.

Any change in the field of energy unavoidably necessitates evaluation by the state in Turkey. The major reason for this is that about 70% of the energy needed in Turkey is provided from foreign countries. Thus, any problematic situations that arise outside the country exacerbate the problems surrounding the condition of domestic resources, whether or not they are sufficiently evaluated, and the country is prepared for the shortages.

Demand for energy increases in line with economic growth. In developing countries, where economic and population growth are generally higher, the rise in energy demand is greater than that of developed countries. Turkey's energy demand is increasing by approximately 5.5-5.6% on average. However, in recent years, this rate has sometimes exceeded 8%.

Table 1 shows the energy consumption of Turkey and projections for the future. The primary energy consumption of Turkey was equivalent to about 80 million tons of oil energy in 2000, about 130 million tons of oil energy (mtoe) in 2005 and is expected to amount to the equivalent of 300 million tons of petrol energy in 2020.

Table 1: Total Final Energy Consumption in Turkey (mtoe)

Energy Sources	**1990**	**2000**	**2005**	**2010**	**2020**
Coal and lignite	16.94	23.32	35.46	39.70	107.57
Oil	23.61	31.08	40.01	51.17	71.89
Gas	2.86	12.63	42.21	49.58	74.51
Com. Renewables and Wastes	7.21	6.56	5.33	4.42	3.93
Nuclear	–	–	–	–	7.30
Hydropower	1.99	2.66	4.16	5.34	10.00
Geothermal	0.43	0.68	1.89	0.97	1.71
Solar/Wind/Other	0.03	0.27	0.22	1.05	2.27
Total Primary Energy Consumption	53.01	77.49	129.63	152.22	279.18

Source: Ibrahim Yüksel (2008). "Global Warming and Renewable Energy Sources for Sustainable Development in Turkey", Renewable Energy, 33: 802-812, www.elsevier.com/locate/renene/10.01.2011

Projections by the Turkish Ministry of Energy and Natural Resources (TRMENR) put Turkish energy consumption at 300 mtep per year by 2020. Current energy policy foresees as large as possible an increase in the rate of domestic and renewable energy resources and a reduction of foreign dependency by 3% over the coming decade in order to meet this demand in 2020.[2]

2 Yusuf Yazar (2010). "Status and Future of Turkey's Energy", **Seta Analysis,** Vol. 31, www.setav.org/ 25.12.2010.

As the aforementioned clearly demonstrates, Turkey is a net energy importing country and is dependent on foreign sources.

2. Energy Production

Energy is provided from two major resources; fossil fuels and renewable resources. For example, in the USA and some other industrialized countries, most of the demand for energy is satisfied from fossil fuels, yet the questions of how that energy is generated, and how it is harming the environment, are little considered. Attention focuses on the security of energy supply.[3]

The existing energy policy of Turkey aims to provide economic and safe energy supply in order to support economic growth, and to improve and support social development programmes, without endangering the environment. TRMENR assesses the energy needs of the country and decides on the best strategy decisions for its energy policies in the short and long term. The government, private sector, and foreign investors must increase their investments in domestic energy resources to rapidly improve efficiency.[4]

Table 2: Total Final Energy Production in Turkey (mtoe)

Energy Sources	1990	2000	2005	2010	2020
Coal and lignite	12.41	13.29	20.69	26.15	32.36
Oil	3.61	2.73	1.66	1.13	0.49
Gas	0.18	0.53	0.16	0.17	0.14
Com. Renewables and Wastes	7.21	6.56	5.33	4.42	3.93
Nuclear	–	–	–	–	7.30
Hydropower	1.99	2.66	4.16	5.34	10.00
Geothermal	0.43	0.68	0.70	0.98	1.71
Solar/Wind/Other	0.03	0.27	0.22	1.05	2.27
Total Primary Energy Consumption	25.86	26.71	34.12	39.22	58.20

Source: İbrahim Yüksel (2008). "Global Warming and Renewable Energy Sources for Sustainable Development in Turkey", Renewable Energy, 33.

In 2005, primary energy production amounted to 34.12 million tons of oil equivalent (mtoe). The most important developments in the country were achieved in the production of hydropower, geothermal energy, solar energy, and

3 Howard Geller (2002). "Energy Revolution: Policies for a Sustainable Future", Island Press, Washington DC.

4 TRMENR, Energy Report, 1996.

coal. Since 1990, the use of hydropower, geothermal, and solar energy in Turkey has grown. However, due to the increased role of natural gas in the system and decreased use of non-tradable biomass, the share of total renewable resources in the total primary energy supply has declined.[5]

TRMENR is the highest authority for developing and implementing the general and, particularly, the renewable energy resources. The Code on the use of Renewable Energy Resources was adopted for the production of electrical energy in May 2005. The objectives of this code are to increase the use of renewable energy resources for generating electrical energy, to diversify the energy resources, to decrease greenhouse gas emissions, to evaluate waste, and to protect the environment. The production, consumption, and transformation of energy have an important effect in increasing greenhouse gas and air pollution. Energy production must be cheap, safe, ecological, renewable, suitable, and economic. Project drafts must be developed that promote energy savings.[6]

3. Dependency on Energy

Since alternative energy resources are nature-based, they reduce the need to invest in energy imports and decrease foreign debts. The necessary energies are easy to use and there is no need for specialists (Mutlu, 2002: 66). Due to the fact that they are domestic, they significantly help to reduce dependency on energy imports and to promote employment.[7]

Dependency on imported fuels constitutes a considerable cost factor for countries. African countries spend 80% of their export income on oil imports. Alternative resources, apart from biomass, do not, however, incur any fuel expenditure, and potential price fluctuations do not pose any risk for these energies. In 1994, Spain was prompted by its high unemployment and economic problems to invest in alternative resources. Wind energy provided employment for about 40,000 people.[8]

It is impossible for Turkey to meet its energy consumption needs from domestic sources. Domestic production insufficiency as calculated by the Ministry

---

5 İbrahim Yüksel (2008). "Global Warming and Renewable Energy Sources for Sustainable Development in Turkey", **Renewable Energy 33,** 802-812, www.elsevier.com/locate/renene/ 10.01.2011.

6 E. Zengin, and Ö. F. Ünal (2010) "Turkey's Energy Sector and Some Remedies Against Environmental Pollution", http://www.meps10.pwr.wroc.pl/submission/data/papers/P58.pdf 15.12.2010.

7 www.deltur.cec.eu.int 10.01.2011.

8 Janet Savin (2003). "Creating A New Energy Future", **State of the World 2003,** Turkish version by Şehnaz Tahir Gürçağlar, TEMA Foundation Publications, İstanbul, 103-135.

of Energy and Natural Resources is shown in Tables 3 and 4. While 78.4 mtp of energy were consumed in 2002, production in the same year amounted to 24.6 mtp. Thus, production meets 31% of demand. Energy production in 2010 is expected to meet 29% of demand. In 2020, it is expected that 30% of demand will be met by domestic production.

Table 3: Primary Energy Balance of Turkey and Its Development (mtpe)

	2002	2010	2020
Demand of Energy	78.4	126	222
Domestic Production	24.6	37	66

Source: Turkish Ministry of Energy and Natural Resources

Of the total imports into the Turkish economy, the manufacturing industry is followed in second place by energy imports. For example, in 2007, energy imports accounted for about 20% of total imports. Table 4 compares the development of Turkish energy imports to total imports over the period from 1980-2007. Energy imports totalled $ 3.907 million in 1980, but increased to $ 33.833 million in 2007. 47% of the foreign trade deficit resulted from energy imports, *and energy imports more than tripled. Energy imports decreased in the wake of the crisis in 2001. According to the economic indicators in 2007, nearly half of Turkey's trade deficit resulted from energy imports.*[9]

Table 4: Turkish Energy Imports and Exports (million dollars)

Year	Energy Imports	Energy Exports	Energy Imports/ Total Imports (%)	Net Energy Imports/ Balance of Trade (%)
1980	3.907	59	49.4	77.0
1985	3.780	377	33.3	100.5
1990	4.622	296	20.7	46.3
1995	4.621	290	12.9	30.8
2000	9.541	329	17.5	34.5
2001	8.339	445	20.1	78.4
2002	9.204	692	17.9	54.9
2003	11.575	980	16.7	48.0
2004	14.407	1.429	14.8	37.8
2005	21.256	2.641	18.2	43.0
2006	28.859	3.567	20.7	46.8
2007	33.833	5.148	19.9	45.7

Source: Turkish Ministry of Energy and Natural Resources

9 Treasury, *2008.*

Turkey's primary energy production is not sufficient to meet the country's energy needs. In order to meet its growing energy demand, Turkey needs to import natural gas from other countries. This situation makes Turkey completely dependent on oil and natural gas imports.[10]

4. Energy Use and Externality

Over the last two hundred years, the need for fossil fuels has continuously increased due to developments in technology. The energy age, based on the dominance of oil and coal, constituted an environment of insecurity as a consequence of the oil crisis that arose in 1973. As a result of this insecurity, renewable energy resources have attracted new and intense attention all over the world. Thus, "energy security" and "energy diversity" concepts became the determinative elements of energy policy after the crisis.[11]

As the populations in third world countries increase, their current energy requirements are expected to grow by between 50-60%, at least[12]. Studies have shown that the reserves of fossil fuels have begun to decrease, and they cannot respond to the continuously increasing energy demand. According to the figures from 1996, it is estimated that coal will run out after 235 years, oil after 43 years, and natural gas after 66 years. The world is consuming energy 300 thousand times faster than fossil fuels can form. In other words, the fossil fuel that formed over one thousand years is consumed in a short time period.[13] As a result, fossil fuel reserves are depleting and will inevitably be unable to meet the continuously growing energy demand. In the policies of the new world order, two issues have gained importance. The first is technology, and the second is energy. Developing countries are expected to have sufficient and cheap energy resources.[14]

---

10 Havva Çaha (2006). "Energy Security of Turkey", **International Conference on Human and Economic Resources,** İzmir, 83-88.

11 M. Kemal Büyükmihci (2003). "Renewable Energy Sources and Applications of European Union Countries Prepared by the Ministry of Energy and Natural Resources, which measures are planned within the Framework of the Draft Law", **Symposium on New and Renewable Energy Sources,** TMMOB, 3-4 October Kayseri, 15-22.

12 Avijit Gupta (1993). "Turkey in the Third World Environment and Development", Turkish version by Şükrü Alpagut, Kabalci Publications, İstanbul.

13 İlker Yilmaz, Mustafa İlbaş, and Şükrü Su (2003). "Evaluation of Wind Energy Potential in Turkey", New and Renewable Energy Sources Symposium, **TMMOB,** 3-4 October 2003, Kayseri, 399-401.

14 İzzettin Önder (2001). "New World Order and the Energy Policy", **Turkey Third Energy Symposium,** TMMOB, Ankara, 8-18.

As energy consumption increases, so do fuel and natural gas imports. Increasing energy consumption also increases air pollution. When compared to less developed and developed European economies, the carbon emission per capita in Turkey has increased.[15]

Nowadays, Turkish energy policy gives importance and priority to electricity generation from renewable energy resources and to increased efficiency in energy usage, from generation to consumption. In addition, it emphasizes the importance of diversifying energy sources. Due to its geopolitical importance, Turkey places special importance on its role as an "energy corridor" and market liberalization.[16]

5. Renewable Energy Resources and Their Use

The use of renewable energy resources is vitally important for securing the future of all humanity. The amount of carbon dioxide gas released into the atmosphere is causing intensive global warming. Increases in the average temperature will result in global climate change, rising sea levels, and consequently, many important agricultural areas will be submerged under water. The first condition for preventing global warming is to make the energy infrastructure suitable for using renewable energies, and thus reducing the use of fossil fuel.[17]

Renewable energy resources offer more advantages than fossil fuels in that their quantities are not limited, they cause less harm to the environment, and they are safer.[18] The fastest growing and best-known energy resources are solar and wind energies,[19] both of which support further sustainable development.

In a broader sense, renewable energy resources can be considered an energy system that continuously renews itself and causes less pollution. Unlike fossil fuels, these resources do not contain CO_2. Solar, wind, biomass, geothermal and hydro energy are the major kinds of renewable energy resources.

The fundamental advantage of renewable energy resources is that they are available in every part of the earth, depending on their geographical and geopolitical conditions. In other words, these are natural energy resources, and there is no need to import them. These resources reduce dependency on energy imports.[20]

---

15 EIA (2010), http://www.eia.doe.gov/countries/country-data.cfm?fips=TU#data/03.12.2010

16 Yusuf Yazar (2010). "Status and Future of Turkey's Energy", **Seta Analysis,** Volume: 31, www.setav.org/ 25.12.2010

17 Ruşen Keleş, and Can Hamamci (2002). "Ecology", Ankara: İmge.

18 Mutlu, Ahmet (2002). "Nuclear Did Fustiness, Sustainable Energy Did It?", **Standart, Temmuz-2002,** 66.

19 Janet Savin (2003). "Creating A New Energy Future", **State of the World 2003,** Turkish version by Şehnaz Tahir Gürçağlar, TEMA Foundation Publications, İstanbul, 103-135.

20 E. Kepenek (2010). "Renewable Energy: Ideas for Developing Countries", www.stps.metu.edu.tr/stpswp/series06/0611.pdf 18.12.2010.

Table 5: Main Renewable Energy Sources and Their Usage Forms

Energy Source	Energy Conversion and Usage Options
Hydropower	Power generation
Modern Biomass	Heat and power generation, pyrolysis, gasification
Geothermal	Urban heating, power generation, hydrothermal
Solar	Solar home systems, solar dryers, solar cookers
Direct Solar	Photovoltaics, thermal power generation, water heaters
Wind	Power generation, wind generators, windmills, water pumps
Wave	Numerous designs
Tidal	Barrage, tidal streams

Source: Ayhan Demirbaş (2006). "Global Renewable Energy Resources", Energy Sources, Part A, 28.

Table 5 shows how renewable energy resources can be transformed and used. Hydropower is used in the generation of power, and modern biomass is used in heat and power generation, purification, and fragmentation. While geothermal energy is used in urban heating and power generation, wind energy is used in power generation in wind generators, windmills, and water pumps.

Table 6: Turkey's Renewable Energy Potential

Energy type	Intended usage	Natural capacity	Technical	Economical
Solar energy	Electric (billion kWh) Thermal (mtoe)	977 000 80 000	6 105 500	305 25
Hydro power	Electric (billion kWh)	430	215	124.5
Wind Direct energy (land) Direct energy (offshore) Wave energy	 Electric (billion kWh) Electric (billion kWh) (billion kWh)	 400 – 150	 110 180 18	 50 – –
Geothermal energy	Electric (billion kWh) Thermal (mtoe)	– 31 500	– 7 500	1.4 2 843
Biomass energy	Total (mtoe)	120	50	32

Source: Ayhan Demirbaş (2006). "Global Renewable Energy Resources", Energy Sources, Part A, 28

Table 6 illustrates the potential offered by Turkey's renewable energy resources. The potential seems to be quite extensive and constitutes the second largest domestic energy resource after coal. The priority sequence of renewable energy resources in Turkey is: hydro energy, biofuel energy, wind, biogas, geothermal energy, and solar energy.[21]

21 Ahmet Koyun (2007). "Mediterranean and National Strategies for Sustainable Development Priority Field of Action 2: Energy and Climate Change Energy Efficiency and Renewable Energy

Turkey imports more than half of its energy need from different countries. Natural gas accounts for the largest share of these imports, and has been imported since 1987. Because of its geographical advantage, Turkey's potential for solar, wind, hydro, and geothermal energies is particularly high.[22]

Turkey does not have large oil and gas resources. The fundamental internal energy resources are lignite, hydro and biomass. Thus, Turkey must develop new long-term energy strategies, in order to decrease the importance of fossil fuels in its primary energy consumption. The use, development, and technologies of renewable energy resources are becoming increasingly important for sustainable economic development in Turkey.[23]

Turkey has a potentially relatively rich diversity of renewable energy resources. Turkey is the seventh richest country in the world in terms of potential geothermal direct use and electricity generation.[24] Despite the enormous renewable energy potential in Turkey, its share in total energy consumption decreased in 2007 to 4.5%. Energy production from renewable energy resources largely remained constant between 2000 and 2007.[25]

Thanks to its renewable energy resources, Turkey can potentially meet its energy supply and consumption needs in the future. In order to increase this potential, Turkey has no choice but to start redeveloping and implementing the renewable energy resources. The application of renewable energy technologies decreases both existing global and environmental problems, and the national energy insecurity associated with the use and production of fossil fuels.[26]

The reserves of fossil energy resources, such as oil, natural gas, and coal, are continuously decreasing, and the danger of environmental pollution is prompting the world to focus on renewable energy resources. According to World Energy Agency statistics, investments in renewable energy resources will reach 10 million dollars by 2030. In 2020, energy demand will be 400-500 billion kWh, of which 200 billion kWh will be generated from Turkey's respective resources. In this respect, adequate steps need to be taken in order to lessen dependency on energy imports.

Turkey", **National Study's Summary**, http://www.planbleu.org/publications/atelier_energie/TR_Summary.pdf/ 10.01.2011.

22 E. Kepenek (2010). "Renewable Energy: Ideas for Developing Countries", www.stps.metu.edu.tr/stpswp/series06/0611.pdf/ 18.12.2010.

23 Dilek Temiz, and Aytaç Gökmen (2010). "The Importance of Renewable Energy Sources in Turkey, **International Journal of Economics and Finance Studies**, Vol. 2, No. 2, http://www.sobiad.org/eJOURNALS/journal_IJEF/archieves/2010_2/03dilek_temiz.pdf/ 30.12.2010.

24 F. Demir (2002). "A Failure Story: Politics, Society And Financial Liberalization in Turkey: The Paths of Retransformation in The Post-Liberalization Era", http://www.econturk.org/ 10.12.2010.

25 D. B. Budak (2009). "Analysis of Renewable Energy and its Impact on Rural Development in Turkey", www.agripolicy.net/ 25.12.2010.

26 E. Zengin, and Ö. F. Ünal (2010) "Turkey's Energy Sector and Some Remedies Against Environmental pollution", http://www.meps10.pwr.wroc.pl/submission/data/papers/P58.pdf/ 15.12.2010.

Not only are the amounts of energy produced from the sun, wind, isothermals, and sea waves not sufficient, they are also uncontrollable.[27] The enormous amount of capital investment in the installation of alternative energy resources is the biggest hurdle to these energy resources. The situation is also influenced by market activity and lending costs. The German government, for example, granted cheap loans, which were provided by major banks and backed by the federal government, and offered tax cuts. Over time, billions of dollars have been invested in wind energy. The increased number of jobs ensured a large political support base behind the sector.[28]

Table 7: *Installed Power and Production of Electricity Plants (1990-2007)*

	Installed Power (MW)					Production (GWh)				
Years	**Thermal**	**Hydro**	**Geothermal and Wind**	**Total**	**Change (%)**	**Thermal**	**Hydro**	**Geothermal and Wind**	**Total**	**Change (%)**
1990	9535.8	6764.3	17.5	16317.6	3.2	34314.9	23148.0	80.1	**57543.0**	10.6
1995	11074.0	9862.8	17.5	20954.3	0.5	50620.5	35540.9	86.0	**86247.4**	10.1
2000	16052.5	11175.2	36.4	27264.1	4.4	93934.2	30878.5	108.9	**124921.6**	7.3
2001	16623.1	11672.9	36.4	28332.4	3.9	98562.8	24009.9	152.0	**122724.7**	-1.8
2002	19568.5	12240.9	36.4	31845.8	12.4	95563.1	33683.8	152.6	**129399.5**	5.4
2003	22974.4	12578.7	33.9	35587.0	11.7	105101.0	35329.5	150.0	**140580.5**	8.6
2004	24144.7	12645.4	33.9	36824.0	3.5	104463.7	46083.7	150.9	**150698.3**	7.2
2005	25902.3	12906.1	35.	38843.5	5.5	122242.3	39560.5	153.4	**161956.2**	7.5
2006	27420.2	13062.7	81.9	40564.8	4.4	131835.1	44244.2	220.5	**176299.8**	8.9
2007	27271.6	13394.9	169.2	40835.7	0.7	155196.2	35850.8	511.1	**191558.1**	8.7

Source: TEIAS

The initial investments in renewable energy resources are expensive relative to the other resources, but the long run returns would be sufficient enough for the national economy. In the first step, the costs could be reduced through governmental support. On the other hand, the incentives provided to support attractive

27 Biltekin Özdemir (2009). "Global Pollution Sustainable Economic Growth and Environmental Taxes",**Journal of Finance**, Volume:156, January-June, 1-36.

28 Janet Savin (2003). "Creating A New Energy Future", **State of the World 2003,** Trs: Şehnaz Tahir Gürçağlar, TEMA Foundation Publications, İstanbul, 103-135

possibilities for local and foreign investors must protect the interests of the country, and be able to adapt to the changing conditions.[29]

By European comparison, Turkey is a country with relatively high potential in terms of hydro, wind, and geothermal energies. The contribution of alternative energy resources other than biomass to satisfying energy demand is trending upwards.[30] Table 7, indicates the installed power and production of electricity plants for the period from 1990 – 2007. The total installed capacity of 16,318 MW in 1990 reached 40,835 MW in 2007. Electricity generated from thermal, geothermal and wind plants was 34,315 GWh, 23,148 GWh and 80.1 GWh respectively in 1990, while they increased to 155,196 GWh, 35,851 GWh, and 511 in 2007. Apart from 2001, annual production sometimes increased more than 10% in this period.

6. Renewable Energy and Tax

A lot of the carbon dioxide emissions that cause the greenhouse effect result from the use of fossil fuels in energy consumption. Therefore, a tax policy that encourages the reduced use of fossil fuel and the use of renewable energy resources that do not harm the environment instead of fossil fuels, will contribute towards decreasing environmental externalities.[31]

Nowadays, as globalization progresses, both all countries individually and the world as a whole are facing a large number of common problems which cannot be solved alone. These include environmental pollution of all continents, the atmosphere and the world seas, depletion of the ozone layer, global warming and climate change, disappearance of animal and plant diversities, acid rain, use of water, waste management, global poverty, financial instability, epidemic diseases, wars, international environmental refugees, urbanization and noise. There is no doubt that preventing and decreasing the effects, harm and losses caused by these negative externalities can only be overcome by solutions and measures that require global cooperation, international agreements, solidarity and world-wide approaches.

29 Kamil Kaygusuz, and Ahmet Sari (2003). "Turkey's Current Energy Situation, Sustainable Development and Renewable Energy Sources", **New and Renewable Energy Sources Symposium**, TMMOB, 3-4 October 2003, Kayseri, 347-356.

30 Bengü Z. Erdem (2010). "The Contribution of Renewable Resources in Meeting Turkey's Energy-Related Challenges", **Renewable and Sustainable Energy Reviews 14**, 2710-2722.

31 Bernard P Herber, and Jose T. Raga (1995). "An International Carbon Tax to Combat Global Warming: An Economic and Political Analysis of the European Union Proposal", **American Journal of Economics and Sociology**, Vol. 54, No. 3, 257-267.

Taking environmental costs into account industrial corporations complain about the excessive regulations and control tools. These costs are not the matter of dispute theoretically, but they affect competitiveness. Since natural resources are determined by global market prices, environment-related taxes should be levied globally. Specific solutions should be developed for each environmental problem. The costs arising from negative externalities on the environment – a classic public property – must be transferred to those individuals or enterprises that use air, water and land for their economic activities and consumption expenditures. Environment taxes, as an alternative economic and financial approach, have been considered a much more effective tool of taxation that is cost-effective, flexible and dynamic, which generates revenue and contributes towards income distribution, generating double dividends by both removing economic externalities and helping to reduce the tax burden on the other tax bases.[32]

Tax is the major public tool for preventing negative externalities. The idea of imposing tax on goods that cause the externality was first put forward by A. C. Pigon. These kinds of taxes are therefore called "Pigon Taxes". C. Plott characterized these kinds of taxes as "corrective taxes" and examined the effects of regulative taxes on the goods that create negative externality.[33]

Pigon taxes guarantee to directly impact the activities of pollutants, increasing the price of pollution to reflect the social cost. These taxes can help to decrease the cost of environmental policy. This approach is described as "cost effective". In addition, these taxes are under consideration as a source of income, which can be reserved for environmental protection and allocated to the general governmental budget. In light of these attributes, Pigon taxes are regarded as the most effective tax criterion in decreasing pollution in line with the theory of externality.[34]

The tax used for controlling pollution is called green tax. As part of its new sustainable environmental policy in 1992, the European Union raised the idea of a green tax on carbon emissions.[35] Taxing carbon dioxide emissions or collecting energy taxes have been accepted as effective instruments in some European

---

32 Biltekin Özdemir (2009). "Global Pollution Sustainable Economic Growth and Environmental Taxes", **Journal of Finance**, Volume:156, January-June, 1-36.

33 Charles R. Plott (1966). "Externalities and Corrective Taxes", **Economica**, New Series, Vol. 33, No. 129, 84-87.

34 Ronald T. McMorran, and David C. L. Nellor (1994). "Tax Policy and the Environment: Theory and Practice", International Monetary Fund, Working Paper No. 94-106.

35 Michael Parkin, Melanie Powell and Kent Matthews (1997). "Economics" (3rd. Edition), England: Addision Wesley.

countries, where both energy tax and carbon dioxide tax are collected on the basis of energy consumption or carbon dioxide emission.[36]

In Turkey, fuel consumption tax, motor vehicle tax and environment tax are considered green taxes. However, the fact that these taxes aim primarily to generate financial benefit and only a small part of the income is used to protect the environment does hinder their evolution into a tool for directing and controlling as is the case in EU countries. As part of Turkey's adaptation to the Acquis Communautaire, Turkish environmental taxes should be directed at economic activities that cause environmental damage. The use of technologies that damage the environment less should be encouraged. As a result of this and in an attempt to protect the environment, the level of environmental pollution can be decreased, and economic development can be encouraged by lowering the taxes on manpower and capital.

In Turkey, the primary reason for taxing energy is fiscal, and far from the aim of protecting the environment. The related taxes are easily collected, and provide a considerable amount of revenue. In addition, tax policies play an important role in renewable energy generation and consumption. Tax incentives could help to increase production, thereby decreasing the demand for fossil fuels, which pollute the environment and are expensive.

Conclusion

Fossil fuels, such as oil and coal, are the most important energy resources, but are rapidly depleting. Greenhouse emissions released during the processes of energy production and consumption are among the main causes of global warming and climate change. Ensuring energy supply security, decreasing our dependency on foreign energy from its current level of 73% and eliminating the associated risks, increasing the efficiency of efforts to halt climate change across the entire process from production to consumption of energy, preventing waste, and decreasing energy intensity are enormously important.

In Turkey, laws have recently come into effect that will play an important role, particularly in respect of controlling consumption-induced greenhouse gases. Codes and regulations incorporate the use of renewable energy resources in electricity production, govern the more efficient and economical consumption of energy, and ban the use of nuclear energy in electricity production.

---

36 Xuejun Wang (1998). "Taxation Policy: Its Role in Environmental Protection and Resource Conservation in China", **Center for Sustainable Resource Development**, University of California, Berkeley.

Turkey is a rich country in terms of its numerous renewable energy resources. It is especially superior in the areas of hydro, wind, solar, and geothermal energy. These resources need to be put to efficient use, now. Raising individual and corporate awareness in respect of producing and consuming renewable energy resources is not enough; the government must expand its legal and tax regulations, in order to encourage the production and consumption of these resources.

In terms of regulating the energy prices, taking environmental outcomes of using fossil fuels into consideration, energy prices will be important policy instruments. But, this regulation on its own is not enough. The lack of information and barriers for accessing the market in some industries prevent the new technologies and applications from functioning.

It is necessary to gain the support of central and local governments and nongovernmental organizations for substituting 50% non-renewable energies with renewable energies. At the same time, individual and social awareness needs to be raised to change lifestyles in favour of this development (Atagündüz, 2001: 262).

Despite increasing energy prices, worries about global warming and climate change, and the rising demand for energy, dependency on – slowly depleting – fossil fuel dependency will continue into the future; developments in the areas of new technologies are still far from commercial maturity and cannot yet meet the increasing demand. The concerns regarding energy security can be summarized as follows:

- Energy should be accessible in terms of cost, time, and quantity.
- Public and private sectors are driven by free market applications.
- Dependency on foreign sources needs to be reduced.
- Regional/international effectiveness of a country is relative to its potential in energy.
- Resource and technology diversity should be aimed for.
- Usage of renewable resources has to be increased to the maximum possible.
- Efficiency in both producing and using energy should be increased.
- The impacts on the environment have to be lowered to a minimum, in both the production and usage of energy and natural resources.

Turkey needs to proceed rapidly to a new energy paradigm, substituting its energy supply with renewable and local resources. Its renewable energy resources could potentially make an enormous contribution to its future sustainable energy independence. This approach will help Turkey to realize its environmental targets, decrease its supply dependency, and increase its energy security. The renewable energy potential that exists in Turkey must be tapped with the help of rational support and government incentives.

References

Atagündüz, Gürbüz (2001). “Measures to Slow Down some of the World Climate Models and Climate Change Speed”, **Turkey Third Energy Symposium**, “Globalization” in the Energy Sector Structural Change Programme and the National Energy Policy, 5-6-7 December, Ankara, 253-262.

Budak, D. B. (2009). “Analysis of Renewable Energy and its Impact on Rural Development in Turkey”,www.agripolicy.net/ 25.12.2010

Büyükmihci, M. Kemal (2003). “Renewable Energy Sources and Applications of European Union Countries Prepared by the Ministry of Energy and Natural Resources, which measures planned within the Framework of the Draft Law”, **Symposium on New and Renewable Energy Sources,** TMMOB, 3-4 October Kayseri, 15-22.

Çaha, Havva (2006). “Energy Security of Turkey”, **International Conference on Human and Economic Resources,** İzmir, 83-88.

Demir, F. (2002). “A Failure Story: Politics, Society And Financial Liberalization in Turkey: The Paths of Retransformation in The Post-Liberalization Era”, http://www. econturk.org/ 10.12.2010.

Demirbaş, Ayhan (2006). “Global Renewable Energy Resources”, **Energy Sources, Taylor and Francis Ltd,** Volume: 28: Number: 8, 779-792.

Demirbaş, Ayhan (2006). “Turkey’s Renewable Energy Policy”, **Energy Sources, Taylor and Francis Ltd,** Volume:28, Number: 7, 657-665(9)

EIA (2010), http://www.eia.doe.gov/countries/country-data.cfm?fips=TU#data/ 03.12.2010.

Erdem, Bengü Z. (2010). “The Contribution of Renewable Resources in Meeting Turkey’s Energy-Related Challenges”, **Renewable and Sustainable Energy Reviews 14,** 2710-2722.

Geller, Howard (2002). “Energy Revolution: Policies for a Sustainable Future”, Island Press, Washington DC.

Gupta, Avijit (1993). “Turkey in the Third World Environment and Development” (Turkish version by Şükrü Alpagut), Kabalci Publications, İstanbul.

Herber, Bernard p. , and Jose T. Raga (1995). “An International Carbon Tax to Combat Global Warming: An Economic and Political Analysis of the European Union Proposal”, **American Journal of Economics and Sociology**, Vol. 54, No. 3, 257-267.

Kepenek, E. (2010). “Renewable Energy: Ideas for Developing Countries”, www.stps.metu. edu.tr/stpswp/series06/0611.pdf/ 18.12.2010.

Kaygusuz, Kamil, and Ahmet, Sari (2003). “Turkey’s Current Energy Situation, Sustainable Development and Renewable Energy Sources”, **New and Renewable Energy Sources Symposium**, TMMOB, 3-4 October 2003, Kayseri, 347-356.

Keleş, Ruşen, and Hamamci, Can (2002). "Ecology", Ankara: İmge
Koyun, Ahmet (2007). "Mediterranean and National Strategies for Sustainable Development Priority Field of Action 2: Energy and Climate Change Energy Efficiency and Renewable Energy Turkey", **National Study's Summary**, http://www.planbleu.org/publications/atelier_energie/TR_Summary.pdf/ 10.01.2011.
McMorran, Ronald T., and David C.L. Nellor (1994). "Tax Policy and the Environment: Theory and Practice", International Monetary Fund, Working Paper No. 94-106.
Mutlu, Ahmet (2002). "Nuclear Did Fustiness, Sustainable Energy Did It?", **Standart, Temmuz-2002,** 66.
Önder, İzzettin (2001). " New World Order and the Energy Policy", **Turkey Third Energy Symposium,** TMMOB, Ankara, 8-18.
Özdemir, Biltekin (2009). "Global Pollution Sustainable Economic Growth and Environmental Taxes", **Journal of Finance**, Vol. 156, January-June, 1-36.
Pamir, Necdet (2008). "Energy Policies of the Republic in Turkey Today", **Turkish Economy,** Ed: Gülen E.ARSLAN, Gazi Bookstore, Ankara. 93-159.
Parkin, Michael, Melanie Powell and Kent Matthews (1997). "Economics" (3rd.Edition), England: Addision Wesley.
Plott, Charles R. (1966). "Externalities and Corrective Taxes", **Economica**, New Series, Vol. 33, No. 129, 84-87.
Savin, Janet (2003). "Creating A New Energy Future", **State of the World 2003,** Trs: Şehnaz Tahir Gürçağlar, TEMA Foundation Publications, İstanbul, 103-135.
Temiz, Dilek, and Aytaç Gökmen (2010). "The Importance of Renewable Energy Sources in Turkey, **International Journal of Economics and Finance Studies**, Vol. 2, No. 2, http://www.sobiad.org/eJOURNALS/journal_IJEF/ archieves/2010_2/03dilek_temiz.pdf/ 30.12.2010.
Wang, Xuejun (1998). "Taxation Policy: Its Role in Environmental Protection and Resource Conservation in China", **Center for Sustainable Resource Development**, University of California, Berkeley.
Yazar, Yusuf (2010). "Status and Future of Turkey's Energy", **Seta Analysis,** Vol. 31, www.setav.org/ 25.12.2010.
Yilmaz, İlker, Mustafa İlbaş, and Şükrü Su (2003). "Evaluation of Wind Energy Potential in Turkey ", New and Renewable Energy Sources Symposium, **TMMOB**, 3-4 October 2003, Kayseri, 399-401.
Yüksel, İbrahim (2008). "Global Warming and Renewable Energy Sources for Sustainable Development in Turkey", **Renewable Energy 33:** 802-812, www.elsevier.com/locate/ renene/ 10.01.2011.
Zengin, E, and Ö. F. Ünal (2010) "Turkey's Energy Sector and Some Remedies Against Environmental pollution", http://www.meps10.pwr.wroc.pl/submis sion/data/papers/P58. pdf/ 15.12.2010.

Enterprise Risk Management for the Energy Industry

Bünyamin Topçu and Şule Güneş

1. Introduction

Recent changes in the New York Stock Exchange (NYSE) Corporate Governance Rules now include explicit requirements for NYSE registrant audit committees to assume specific responsibilities with respect to "risk assessment and risk management" including risks that go beyond financial reporting.[1]

Managers face a variety of new challenges in their day to day business. Globalization, e-business, new organizational partnerships, and the increasing speed of business activity are rapidly changing and expanding the risks faced by organizations. Risk management must now go well beyond insuring against hazards and traditional financial worries, in order to prevent a wide variety of risks concerning strategic, operational, reputational, regulatory and information areas.

As an emerging management discipline, ERM varies across industries and corporations. The insurance industry, financial institutions and the energy industry are among the industries where ERM has seen relatively advanced development in a wide range of corporations. Many government organizations and private companies are using targeted risk management approaches. An increasing number have found that there are significant gains to be made by working systematically based on a commercially oriented overview of risks and opportunities in the short and long terms. Recently, even the public sector has started becoming aware of the potential value of ERM, and risk managers are increasingly bringing it to top executives.[2] Therefore, risk management and ERM have become "hot" subjects in recent years, and many companies and academic programmes have shifted their focus towards this promising new field.

Risk management is one of the issues that is gaining increasing strategic importance in energy and resource companies. In recent years, enterprise risk management has emerged as an important new business trend. This document outlines the practical uses of ERM for energy companies, ones that can provide

1 New York Stock Exchange (NYSE), (2003). Final NYSE Corporate Governance Rules: NYSE, New York.

2 Brockett, P.L. (2010). Enterprise risk management. Encyclopedia of Quantitative Risk Analysis and Assessment: John Wiley & Sons.

organizations with new steps they can take to enhance business decision making and, potentially, shareholder value.

This chapter will help you put in place the management controls and reporting structures necessary to ensure that your hedging programme achieves its goals and does not add unexpected or unwanted risks to your firm.

2. Risk and Risk Management

Risk can be defined as the likelihood that the outcome of a process will not meet expectation.[3]

There are a number of differing types of risks that have varying degrees of impact and therefore need to be treated accordingly.

Fig. 1: Environment of Risk

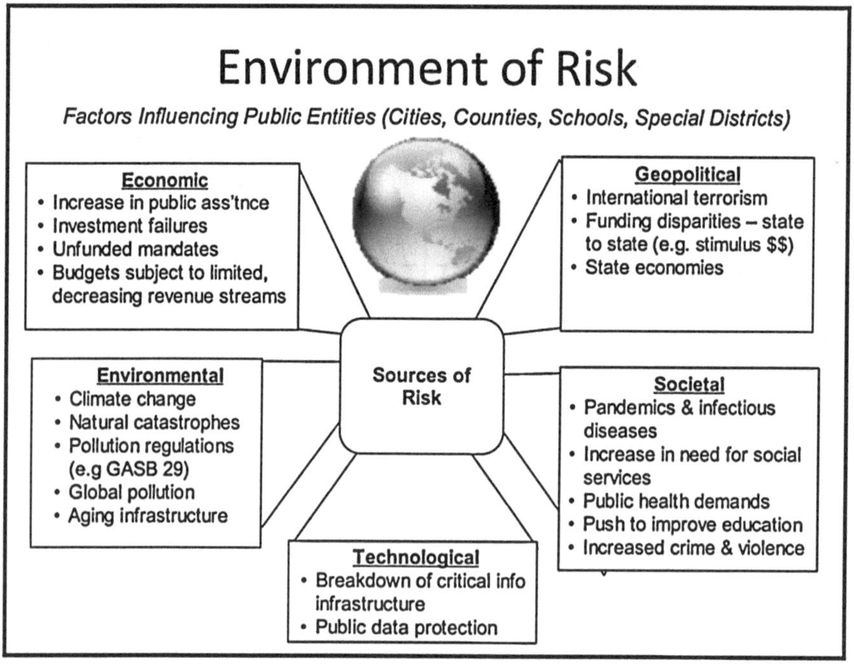

Source: RIMS, (2009). Executive report on enterprise risk management technology.

3 O'Donnell, E. (2005). Enterprise risk management: A systems-thinking framework for the event identification, International Journal of Accounting Information Systems, 6 (3): 177-195.

There are different kinds of risks in each and every industry. Some of them are listed and briefly explained below, and include strategic risks, compliance risks, financial risks, operational risks, environmental risks, employee risks, project risks, political and economic risks.

Strategic risk is the threats to future opportunities and risks that are so significant that they could materially impact the enterprise's achievement of its main purpose or even survival. The biggest risk it presents is the risk that an organization may undertake the wrong operations to achieve its strategic objectives.

Project risk includes all the risks that come with a project being managed separately from the rest of the business or organization.

Operational risk includes all the other risks faced by an operational business. These include the risks, which arise from projects undertaken in the past that are now in their operational phases and are no longer being managed as separate projects but as part of the main operational business.

Risk management is about identifying risk, assessing the impact on your business if a security incident arises, and making the right financial decisions about how to deal with the results of your assessment. It also includes the implementation of a programme to continually measure and assess the effectiveness of existing safeguards in protecting your critical assets.

Managing risk is not a one-time operation; it is an ongoing process.

The first step in the cycle is to conduct a risk assessment, which organizations use in order to determine the extent of potential threats.

The risk assessment is followed by an evaluation of the risks that have been identified. The first evaluation task is prioritizing the various risks. During the prioritization process, the assets with the highest risk levels should be made the highest priority. The final evaluation is carried out to identify safeguard measures and associated costs to reduce each system risk to a non-threatening level.

Following this stage, management selects and implements the appropriate approach to managing each risk. Risk management is based on the premise that risk can probably never be reduced to zero; the cost would be too high. Rather, management's goal should be to address specific risks in one of the following ways:

- Rejecting risk: By rejecting risk, a manager is saying that he doesn't believe the risk exists. So he does nothing. This is the head-in-the-sand approach. Ignore it and maybe it will go away.
- Accepting risk: Accepting risk might seem the same as rejecting risk, but it's actually very different. When a manager performs a security safeguard cost/benefit analysis, he might discover that the cost of mitigating the risk is greater than the probable impact on the business.

- Transferring risk: Transferring risk is most often accomplished by purchasing insurance to cover some or all of the cost of a security incident.
- Mitigating risk: Mitigation involves the implementation of security safeguards across people, processes, and technology in a reasonable and appropriate manner so that the risk is reduced to an acceptable level.

Once the assessment and management steps are complete, you should measure the effectiveness of your risk management efforts. The metrics should be simple and meaningful.[4]

The purpose of risk management is not risk avoidance but knowing what risks are worth taking.

Risk management is moving well beyond the tradition of risk mitigation (using controls to limit exposure problems) towards risk portfolio optimization (determining the organization's risk appetite and capacity among a group of risks across the enterprise, seizing opportunities within those defined parameters, and capitalizing on the rewards that follow). As a consequence, risk management is beginning to be perceived as a new means of strategic business management linking business strategy to everyday risks.[5]

3. Enterprise Risk Management

A recent trend in corporate governance has been the development of an integrated, enterprise-wide approach to assessing business risks that can affect an organization's ability to achieve its business objectives, and to developing programmes for managing those risks.[6]

Definitions of ERM also vary widely, but many agree that it is a top-down approach, based on and supportive of organizational strategy, which is focused on new ways of managing and presenting the risks with the greatest threat to the board and management.[7]

---

4 Olzak, T. (2006), Security risk management. http://adventuresinsecurity.com/blog/2006/02/07/security-risk-management.

5 KPMG, (2001). Enterprise Risk Management: An emerging model for building shareholder value. Australia.

6 Miccolis, J., Hiverly, K. and Merkley, B. (2001). Enterprise risk management: trends and emerging practices. Altamonte Springs, FL: Institute of Internal Auditors Research Foundation.

7 KPMG (2001). Enterprise Risk Management: An emerging model for building shareholder value. Australia.

Corporations across a variety of industries have begun to explore enterprise risk management as a paradigm for reducing volatility in revenues and sustaining competitive advantage. To manage risk on an enterprise-wide basis, organizations must expand their focus beyond traditional concepts of risk to include economic and political factors, along with risks related to reputation, ethics and data integrity.[8] Enterprise risk management can be viewed as a firm's opportunity to manage risk strategically so as to take advantage of value enhancing opportunities. In essence, a missed strategic opportunity can be just as serious as a harmful incident.[9]

ERM allows executives to manage risk more effectively by simulating the impact of future external influences on a business. An accurate representation of the business in an ERM model leads to confidence in the model's use as a true decision-support system. The proven benefits of ERM are:

- Alignment of the strategic aspects of risk with day-to-day operational activities
- Facilitation of more transparency for investors and regulators
- Enhancement of revenue and earnings growth
- Control of risk potential

In preparing for principle-based regulations, companies need to have a strong ERM component.[10]

The Committee of Sponsoring Organizations of the Treadway Commission (COSO) framework describes risk management as an ongoing, enterprise-wide process that involves eight interrelated components.[11] Most corporations adopt the definition of ERM as proposed by the COSO in its 2004 ERM framework. In this framework, ERM is defined as "a process, affected by an entity's board of directors, management and other personnel, applied in strategy setting and across the enterprise, designed to identify potential events that may affect the entity, and manage risk to be within its risk appetite, to provide reasonable assurance regarding the achievement of entity objectives". This definition highlights that ERM reaches the highest level of the organizational structure and is directed by corporations' business strategies. The concept of risk appetite is crucial as risk

8 O'Donnell, E. *(2005)*. Enterprise risk management: A systems-thinking framework for the event identification, *International Journal of Accounting Information Systems*, Volume 6, Issue 3, September, Pages 177-195.

9 Beasley, M. S., Richard C. and Dana, H. (2005). Enterprise Risk Management: An Empirical Analysis of Factors Associated with the Extent of Implementation: Journal of Accounting and Public Policy, 24 (6): 521-531.

10 Doodian, J. J., Knott J. and Rech J. (2008). ERM and other considerations related to a principles-based approach.

11 O'Donnell, E. *(2005)*. Enterprise risk management: A systems-thinking framework for the event identification, *International Journal of Accounting Information Systems*, Volume 6, Issue 3, September, Pages 177-195.

appetite reflects a firm's willingness and ability to take on risks in order to achieve its objectives.[12]

Enterprise Risk Management (ERM) must define, for instance, the overall risk appetite and has to apply it to all parts of a firm. Clear objectives must be defined and communicated throughout the firm. The rating of a firm has to be aligned to the **risk appetite**, and implementation and improvement of the **risk culture** are the most important elements of Enterprise Risk Management. Performance evaluations must be based on portfolio aspects in the context of economic and organizational considerations and not only on an accounting or legal basis. The implementation of the ERM framework supports improves **risk awareness** from the beginning, from strategic to operative and from top management to employees. Each organization's **risk tolerance** is unique and will vary according to organizational culture as well as external factors.

The implementation of an Enterprise Risk Management framework supports and improves risk awareness, from strategic to operative level, and from top management to employees. There is no "one size fits all" approach to developing and implementing an ERM framework.[13]

Fig. 2: Enterprise Risk Management Framework

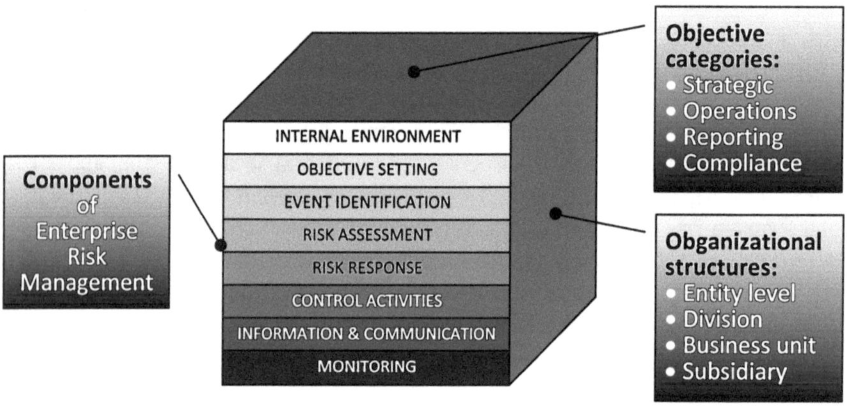

Source: Ciorciari M. and Blattner P. (2008). Enterprise Risk Management maturity-level assessment tool: Chicago: ERM Symposium Publications.

12 Brockett, P. L. (2010). Enterprise Risk Management. Encyclopedia of Quantitative Risk Analysis and Assessment, John Wiley & Sons.

13 Raymond, C. (2008). Guiding principles spell success for ERM: National Underwriter: P&C, 112, (3): 30.

Internal auditors have primary responsibilities related to risk identification and assessment and are likely to interact with senior management on ERM-related implementation issues.

There is extensive academic literature that examines audit quality, but despite presenting some limitations, most of those studies classify the largest international accounting firms, now known as the "Big Four" firms, as high quality auditors. It is possible that organizations committed to carrying out such high quality audits are also more committed to risk management.[14]

4. ERM Compliance and Situation in the World

ERM initially arises from corporations' efforts to comply with laws and regulations. To this end, it is seen more as an efficient internal control process. Within a corporation, it is often conducted with internal control functions and supervised by internal auditors. The most significant regulatory forces that are responsible for the rise of ERM are the Sarbanes Oxley Act of 2002, the Basel Capital Accord II and rating criteria set forth by rating agencies such as Standard & Poor's (S&P).[15]

The crucial point S&P needs to emphasize in its evaluations is not whether management has identified risks according to any specific framework, but rather that a company's management has an embedded and formalized process not only to identify risks, but also to assess, treat, monitor and report the risks it has identified. S&P's greater focus should be on the qualities of the company's risk governance model and the ERM process, not the specific key risk areas.[16]

Complying with Sarbanes Oxley became a costly exercise for many. To ensure compliance, businesses must update and recertify their data on a quarterly and annual basis. The Sarbanes-Oxley Act has also caused many boards of directors and executive teams to look at all the business risks they face: not just financial, but operational, social, ethical, and environmental. These organizations see Sarbanes-Oxley-related expenditure as the beginnings of an ERM platform which will help them meet ever-increasing and shifting regulatory demands.

The Committee of Sponsoring Organizations of the Treadway Commission (COSO) has developed a framework that describes key risk management principles

14 Beasley, M.S., Richard C. and Dana, H. (2005). Enterprise Risk Management: An Empirical Analysis of Factors Associated with the Extent of Implementation: Journal of Accounting and Public Policy, 24 (6): 521-531.

15 Brockett, P.L., (2010). Enterprise Risk Management. Encyclopedia of Quantitative Risk Analysis and Assessment, John Wiley & Sons.

16 RIMS, Risk and Insurance Management Society, (2008). RIMS comments on S&P's Enterprise Risk Management analysis for credit ratings of nonfinancial companies: New York.

Fig. 3: A Short History of ERM

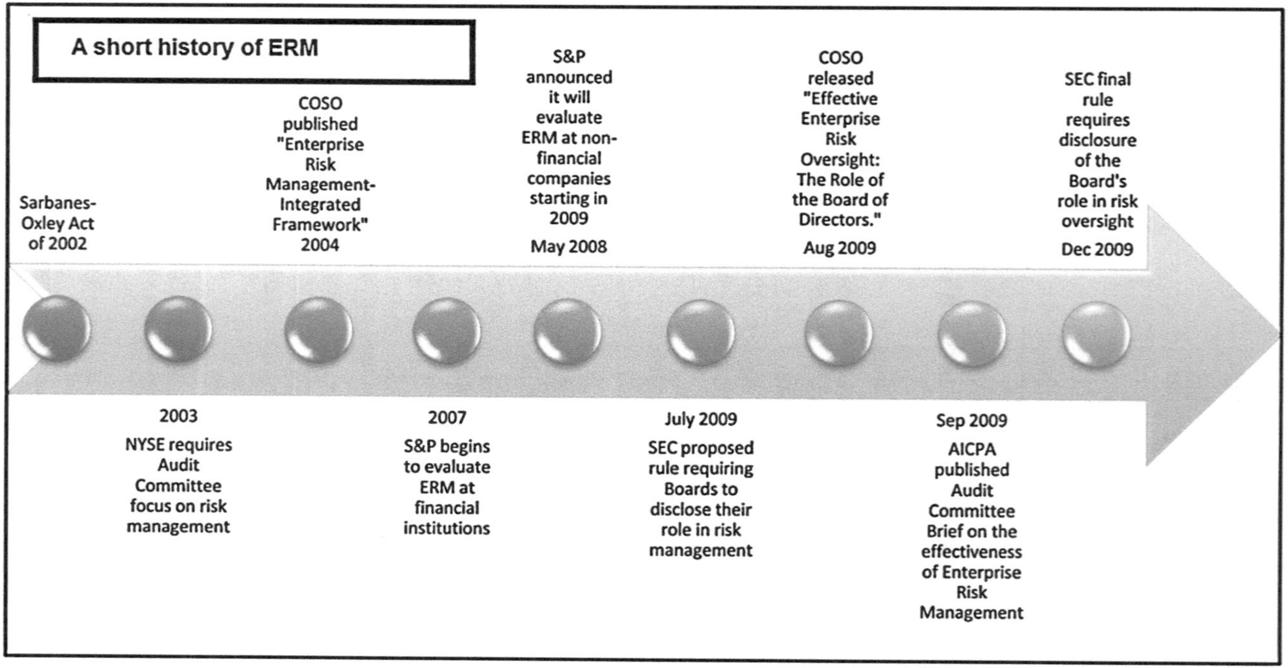

Source: Sander, D. (2010). The risk of enterprise risk management: Sanders Consulting. LLC, Certified Public Accounting Firm.

The Committee of Sponsoring Organizations of the Treadway Commission (COSO) has developed a framework that describes key risk management principles and concepts, provides a common language, and offers direction and guidance. These guidelines build upon and extend the integrated framework for internal control developed by the COSO a decade ago. The original COSO cube has three objectives:

1. Operational objectives, which relate to effective and efficient use of the entity's resources;
2. Financial reporting objectives, which relate to the reliability of all the entity's reporting to internal and external parties;
3. Compliance objectives, which relate to the entity's compliance with applicable laws and regulations (COSO, 2004).

Its five horizontal components consist of: monitoring, information and communications, control activities, risk assessment and control environment.[17]

The recent financial and economic crisis also had an impact on the energy and resources industry. Many companies in the industry have been hit hard by these turbulent times and still face many challenges ahead for the near future.

Deloitte carried out a survey in the energy and resources industry aimed at highlighting how ERM was being implemented in Europe and the Middle East and Africa region. The goal of the survey was to capture and report feedback on the current state of ERM implementation in a cross-section of organizations within the energy and resources industry. The report presents a compilation of the responses, and develops profiles of the leading and prevailing ERM practices in the energy and resources industry.

The survey, "Risk Intelligence in the Energy and Resources Industry," is based on questions and interviews conducted with 50 CEOs, risk managers and senior managements across Europe, the Middle East and Africa.

Internationally, the companies involved in the production of energy also tend to be the frontrunners when it comes to risk management. The reasons for this include the significant risks associated with energy production (such as nuclear reactors), as well as the risks linked to trading of the raw materials used for producing energy, or distributing and selling the energy itself on the international market. Although risk management has not been an integral part of companies that deal with the transportation and distribution of electricity and gas for as long, they are quickly catching up and in certain cases are even forging ahead. Furthermore, water companies are also beginning to take an interest in risk management.

---

17 Matyjewicz, G. and D'arcangelo, J.R. (2005). ERM-based auditing.

The results of this survey mentioned that:

- **Risk management has become general practice in energy and resources companies:** The survey indicates that 96 percent of the companies questioned already conducted risk management operations.
- **Shift from protecting capital equipment to creating value – improving competitive capabilities:** As a company's maturity in risk management increases and these risks are firmly taken control of, other risks relating to strategy and how to implement it, such as the risks associated with the development of new products, penetrating new markets and completing acquisitions, grow in importance.
- Risk reporting in addition to financial reporting – monitoring and reporting risk remains a difficult area: The value of introducing risk reporting alongside traditional financial reporting is becoming more and more recognized by energy and resources companies. However, many of these companies do not yet have any form of integrated risk reporting available.
- **Technology can promote the ERM process; but many companies are still struggling to get to grips with it:** Technology can be used to facilitate the risk management process, but a large number of the organizations surveyed indicated that they were still some distance away from reaching this level.

Thus far, there is still no single solution available that provides the functionalities required to make total ERM possible.[18]

5. Risk Assessment with ERM

The risk assessment must use clear metrics and methodology. Risk mitigation strategies affecting inherent risk must be carried out in order to bring any remaining risks within the risk appetite. Control activities must be based on clear policies and procedures, and focus on objectives, processes and information processing. Relevant information must be communicated, and operations and corrective actions must be monitored on an ongoing basis.

The following main topics relating to risk assessment have been identified and translated into basic ERM elements:

- Event characteristics: When assessing risk, management should consider both the expected and unexpected losses. Correlated events should be assessed together.

18 Deolitte, (2010). Risk intelligence in the energy & resources industry.

- Assessment metrics: Both the possibility of risks occurring and the impact of potential measures that could have an adverse effect on achieving the objectives should be assessed. Risks should be ranked in order of the most threatening so that those can be dealt with first.
- Assessment mode: Management should promote best practices assessment techniques and a continuous and iterative risk management process aligned with the strategy setting process. A composite assessment of risks across the entire board should be performed.

Fig. 4: The Map of Risks

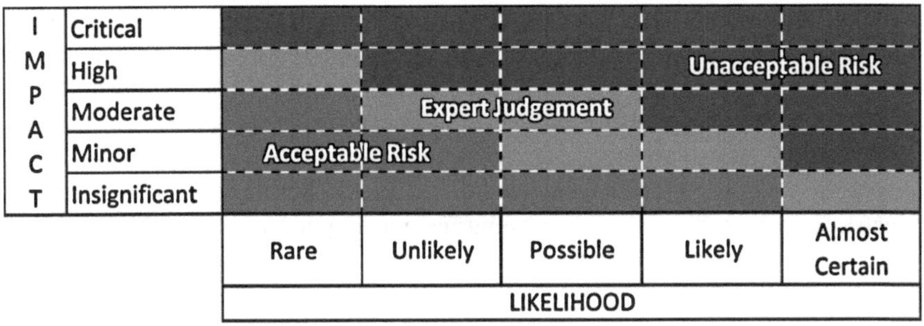

Source: McVey, M. (2007). What is the likelihood and impact of a BA communicating risk?

Management should promote the use of best practice assessment techniques. The quality of the supporting data and any hypothesis should be periodically reviewed. To gauge the quality and accuracy of the applied risk assessment techniques, back-testing should be performed.[19]

6. Energy industry

Developing an efficient energy policy for each country is important. Unfortunately, however, there is no simple and universal response to the problem of resource selection. For instance, let us consider renewable resources like wind, water and solar energy, where opinions differ as to how much they will be able to cope with the potential future energy demands. Oil, gas and coal, which are the only large-scale alternatives, all produce carbon dioxide that contributes to global warming. Although natural gas is twice as efficient and half as polluting as other fossil fuels, dependency on politically unstable countries for this energy

19 Ciorciari M. and Blattner P. (2008). Enterprise Risk Management maturity-level assessment tool. Chicago: ERM Symposium Publications.

source increases the risks to supply with interruption and price instability.[20] Therefore companies operating in the energy industry are faced with many risks both by existing operations and new alternative ones. For example, wind energy protects utilities and energy consumers from the economic risks associated with changing fuel prices, new environmental regulations, uncertain load growth and other cost uncertainties.

Turkey's geographical location makes it an important potential corridor, in particular, for gas and oil from Central Asia and other neighbouring countries to the EU. Turkey's strategic location makes it a natural "energy bridge" between the major oil and gas producing areas in the Middle East and Caspian Sea regions and the consumer markets in Europe.[21]

Continuously supplying cheap electricity of high quality is of critical importance for a number of industries. Dependency on imported energy resources, the effects of increasing fuel prices on cost, and especially changes in unit production costs of electrical energy are vital for supply security. The rate at which resources are used, and also the types of resources chosen especially to produce electrical energy, differ between countries and regions for economic, technical and environmental reasons.[22]

7. Energy Risk Management

Energy deregulation, privatization and competition are a hot international topic. Professionals in this field understand the importance of hedging their financial risk, but are often unclear how to do so. The result is that either they take undue and unwarranted risk or they shy away from futures and derivatives investments that could improve their financial position while preventing substantial losses.[23]

The energy market is the practitioners' guide to trading the markets and optimizing company performance using the correct price risk strategies and tools. Since the turn of the new century, as companies have been exposed to price increases, they have experienced high volatility in energy prices. As a direct trader

20 Topçu, Y.I. and Ülengin, F. (2004). Energy for the future: An integrated decision aid for the case of Turkey Energy. 29, 137-154.

21 Çeliktaş, M. S., Sevgili, T. and Kocar, G. (2009). A snapshot of renewable energy research in Turkey: Renewable Energy, 34 (1): 1479-1486.

22 Akdağ, S.A. and Güler, Ö. (2009). Calculation of wind energy potential and economic analysis by using weibull distribution – a case study from Turkey. Part 1: Determination of Weibull Parameters: Energy Sources, Part B: Economics, Planning, and Policy, 4 (1): 1-8.

23 Fusaro, P.C. (1998). Energy Risk Management: Hedging Strategies and Instruments for the International Energy Markets.

in this market or as an investor in hedge funds in the energy industry, it will give you an insight to the global energy markets and their operation.[24]

Managing risks associated with the energy industry is becoming increasingly complicated due to factors such as government regulations, public policy, financial concerns, and energy resource scarcity. In order to address these issues, affected companies often implement energy risk management strategies. Energy risk management involves the process of pinpointing, evaluating and prioritizing risks associated with uncertainty in the energy market. Controlling energy risks can provide greater market assurance for companies like oil and gas producers, electric providers and gas utilities. It can also give confidence to insurance companies, banks and manufacturers who work with these energy companies.[25]

The security, economic prosperity and social wellbeing of any economy are dependent on a complex system of interdependent infrastructures. Key among these are the energy infrastructure, electric power, oil and natural gas production, transmission, storage and distribution systems – large and small – which fuel and power the economy. Energy markets operate in an environment that is exposed to a variety of risks which attribute to the high volatility of oil, natural gas, electricity and freight prices. The need to control price instability has prompted the development of evaluation and risk management methods for energy assets and their derivatives analogous to those widely used in the fixed income, equity and foreign exchange markets.[26]

Different risks occur in different types of energy markets, and energy risk management strategies may vary depending upon which sector of the energy industry is being assessed. The electricity market, for instance, may pose different threats to a company than the oil or natural gas markets. Other factors that often present energy risks include climate change and environmental concerns.[27]

Many companies, not just financial institutions, are recognizing the need for some kind of integrated risk management focus.

Companies in industries with more quantifiable and hedgeable risks are generally more comfortable discussing ERM, but they tend to focus on controlling specific risks. Examples include: energy, pharmaceuticals, agribusiness and some manufacturers.[28]

---

24 James, T. (2008). Energy markets: price risk management and trading: John Wiley and Sons Ltd.

25 Wisegeek, (2010). What is energy risk management http://www.wisegeek.com/what-is-energy-risk-management.htm.

26 Energy Business Reports, (2008). Risk management in energy markets.

27 Wisegeek, (2010). What is energy risk management. http://www.wisegeek.com/what-is-energy-risk-management.htm.

28 S&P, (2009). Integrating enterprise risk management analysis into corporate credit rating: Progress Report.

In a changing world where energy shortages and climate change have become common concerns, energy and resource companies are being faced with countless new risks. A scarcity of natural resources, political instability, an ageing infrastructure and unpredictable weather conditions are just some of the problems dogging these companies. The oil spill in the Gulf of Mexico is a recent example.

8. Energy Risks

Since the last half of the 19th century, the human lifestyle has relied heavily upon oil and gas. The hunt for oil and gas has brought about many technological advances that have been both beneficial and harmful to the development of mankind. On the one hand, these advances have helped improve the quality of human life and contributed to the reduction of disease, illiteracy, poverty and insecurity. On the other hand, no other source of energy has created such devastating and unabated political instability among nations. The ensuing geopolitical risk identification, management and mitigation with respect to energy sources make up the central question of modern international relations and have ramifications across all boundaries among the mutually dependent economic, political, social and environmental factors that shape the plans and aspirations of nations.[29]

Every form of human activity involves risks of accident or disease, resulting in injury or death. The energy industry is no exception. Although such risks have previously been considered from conventional systems (coal, oil and nuclear), a similar analysis for the so-called alternative or non-conventional systems (solar, wind, ocean thermal and methanol) has been lacking.

The risk of most non-conventional systems, per unit of energy output, is comparable to, and in some cases much higher than, the risks presented by coal and oil. This conclusion is strong if we consider the number of deaths or injuries related to acquiring and producing these energy resources. Nuclear power and natural gas presented the lowest overall risk of the 11 technologies considered, whilst ocean thermal energy ranked third. The most surprising results are that the following eight technologies that were considered were found to be up to 100 times less safe. While a solar collector on the roof appears to be completely innocuous as it silently absorbs sunlight, there is considerable risk inherent in the industrial processes used to construct this device. Per unit of energy output, the solar collector requires a significantly larger input of construction material than any con-

29 MEES, Middle East Economic Survey. (2011). Energy & Geopolitical Risk: MEES Fundamentals.

ventional system. Fabricating and installing the systems, it should be emphasized, also produces risk.

The total risk is calculated by considering six components: material acquisition and construction, emissions caused by material production, operation and maintenance, energy back-up, energy storage and transportation. In this way the risk of widely different systems can be assessed fairly. This methodology of "risk accounting" will not tell us which energy technology to use. However, it can be employed to inform society of the risk inherent in competing energy systems.[30]

Market risk versus business risk: Risk management in the corporate environment is inherently more complex than in a pure financial environment (i.e., trading and investment functions) in that companies have both non-hedgeable business risks (relating to the nature of their specific products and services) and hedgeable market risks (e.g., commodity, currency, interest rate, equity exposures). The level of market risk is furthermore a function of business risks, which can make the implementation of a risk management system a complex process.

9. Energy Risk Management Assessment and Application

An energy risk management assessment typically involves several steps. The first step requires identifying all possible threats posed by energy. Once the potential energy threats have been identified, the risks associated with these threats are usually quantified and prioritized. The final step in an energy risk management assessment usually entails finding ways to address the risk. Solutions may include reducing the energy risk, monitoring the risk or eliminating the risk altogether.

Many companies have expressed an interest in understanding how the principles of Value at Risk (VaR), which were initially developed for managing market risk in a financial environment, can be applied in the corporate environment.

Recent developments in this field emphasize quantitative methods and their application in energy and shipping. Stochastic models for the spot price of oil, natural gas and electricity and of their future curves are reviewed. The pricing of vanilla and exotic derivatives for jump-diffusions is discussed using recent valuation methods based on complex characteristic functions and the Fourier integration. Recent analytical results in stochastic optimal control theory are reviewed and their use is proposed for the solution of a number of complex risk management problems involving portfolios of energy assets and their derivatives.

Applications under discussion include the valuation of storage facilities for natural gas, oil and other fuels, the optimal operation of flexible hydrocarbon

30 Inhaber, H. (2011). Risk of Energy Production: Second Edition.

reservoirs, the hedging of fuel costs in the transportation and energy industries, the valuation of seaborne energy cargoes of oil, products, liquefied natural gas (LNG) transported in tankers, the fuel efficient navigation and weather routing of shipping fleets, the evaluation and hedging of power plants and refineries, the pricing of credit risk and derivative securities in shipping, and the evaluation of electricity storage facilities for offshore wind farms. Further, a checklist has been included to provide some general guidance and a starting point so that energy facilities are able to identify their critical functions and assets, become aware of threats and vulnerabilities, evaluate and rank the threats in terms of the incidents they may cause, and initiate a security enhancement programme, if appropriate.[31]

Many companies have expressed an interest in understanding how the principles of Value-at-Risk, which were initially developed for managing market risk in a financial environment, can be applied in the corporate environment. In our discussions with corporate clients, they have raised a number of issues about implementing a strategic risk management programme within their companies.[32]

9.1 Which hedging strategy to use?

Companies can compare the riskiness of new businesses or different risk profiles. For example, if an energy company is considering setting up a natural gas trading desk, the company can estimate the incremental earnings and earnings volatility that might be incurred. Or, a company considering the use of more floating-rate debt versus fixed-rate debt can estimate the expected interest expense savings and cash flow volatility of different financing alternatives. With these measures, companies can make more informed decisions based on their risk preferences.

9.2 Mitigating risk in volatile markets

Being unaware of compounded risks created in multiple utility data systems, across business units or among key trading partners can substantially alter a company's risk profile, leaving the company exposed to greater market, credit, regulatory and even reputational risks than anticipated. What's needed is a more accurate understanding of net exposure, not only to the risks produced by market dynamics, but also to the significant and often hidden risks and offsetting positions inside the company's operations.

31 Risk Books, (2005). Energy modelling and the management of uncertainty incisive media: 380.

32 Lee, A.Y. (1999). CorporateMetrics Technical Document: RiskMetrics Group.

By delivering a comprehensive energy risk framework with energy commodity risk aggregation and analytics (ECRAA), companies can gain greater confidence in their internal data and also a better understanding of the risks. Energy companies can soften the potential impact of the next worldwide credit crunch, liquidity squeeze or other unforeseen risk events.

Organizations have the ability to proactively capture, analyze and simulate market and credit risks associated with energy commodity price volatility. This front-to-back-office solution includes deal capturing, data integration, advanced predictive analytics, compliance reporting and a full suite of sophisticated risk modelling tools.[33]

Due to the importance of the subject, VaR methodology has been extensively covered in financial literature, especially in recent years. The use of VaR in the energy markets has been described.[34]. It would not be an exaggeration to say that, by now, the methodology, its benefits and shortcomings, as well as various implementation issues are well understood.

VaR is a single, summary, statistical measure of possible portfolio losses. Specifically, VaR is a measure of loss due to "normal" market movements. Losses greater than the VaR are less likely. Subject to the simplifying assumptions used in its calculation, value at risk aggregates all of the risks in a portfolio into a single number that is suitable for use in the boardroom, reporting to regulators or disclosing the information in an annual report.

In order to compute VaR (or any other quantitative measure of market risk), we need to identify the basic market rates and prices that affect the value of the portfolio. These basic market rates and prices are the "market factors." [35]

10. An Enterprise Risk Management Model in the Energy Industry

We developed an enterprise risk management model for energy companies, including direct goals, planning, implementation, risk identification, strategies and also the benefits of ERM for the energy industry.

According to this model, firms should develop a risk management programme with three basic objectives:

33 Http://www.sas.com/offices/europe/turkey/ (11.10.2010).

34 Clewlow, L. and Strickland, C. (2000). Energy derivatives: pricing & risk management.

35 Linsmeier, T.J. and Pearson, N.D. (1996). Risk Measurement: An introduction to value at risk. University of Illinois at Urbana-Champaign.

- Limit the upside price risk of energy product purchases
- Manage (preserve) margins
- Develop a price risk management strategy for the company to use going forward

Created with a thorough understanding of the goals of companies, the model will provide specific hedging strategies designed to meet companies' unique needs.

An energy company model, which recommends several actions to fully implement ERM.

1. Begin with risk inventory and assessment activities.
2. Introduce the ERM framework and methodology in one major operating unit first as a pilot before implementing it throughout the company.
3. Install and make ERM processes operational throughout the company, from the strategic planning process with the board to business processes at operational level.
4. Apply the ERM framework when entering a new business or making other major investment decisions.
5. Reinforce accountability by integrating risk management with individual performance plans in a number of business units.[36]

A risk measurement plan will assist with:

- Developing firms' corporate risk policy.
- Obtaining board approval to initiate a risk management programme
- Establishing a future brokerage account
- Developing relationships with companies that provide non-traditional physical programmes for hedging
- Learning the unique steps to take that reduce price exposure, manage risk, and guard against price spikes.
- Quantifying hedge base percentage.
- Reviewing the structure of supply contracts.
- Assessing the risk profile of the company.
- Establishing an over-the-counter (OTC) account.
- Designing a risk management programme.
- Defining the hedge objectives of the company.
- Determining the appropriate hedge strategies for the company.

---

36 Morgan Stanley Publication, (2006). Corporate risk management: Journal of Applied Corporate Finance, 18 (4).

All energy companies that employ risk management operations gain:

- A detailed hedge programme that fits into the objectives and goals of the company
- A daily mark-to-market report
- A hedge transaction on behalf of the company
- A detailed daily hedge action plan prior to the opening of the New York Mercantile Exchange (NYMEX) futures exchange
- Assistance in negotiations with vendors when purchasing financial risk instruments
- Hedge position management.
- Periodic evaluation of hedge instrument correlations to ascertain that the hedging programme will be compliant with good business practices
- A comprehensive monthly post-mortem of a month's margin protection programme
- Monitoring and adjusting of the hedge strategy.[37]

There is a lot of talk about Enterprise Risk Management delivering broader, more significant benefits to the organization. Some of the benefits of ERM include:

10.1 Energy Risk Management Benefits

We believe corporations can benefit in the following ways by applying risk management concepts at business, senior management, and board of directors' levels:

- Increased transparency of risks: The formalized quantification of the impact of market rate volatility on a company's financial results should significantly improve the clarity of risk awareness within the organization.
- Communication: Risk measures and reporting should improve communication not only between senior management and the business units that assume risk, but also between senior management, the board of directors, shareholders, ratings agencies and regulatory bodies.
- Hedging decisions: Analysis of how risks and expected returns vary for different hedging strategies.
- Capital allocation and performance evaluation: Heightening risk awareness, evaluating the potential impact of adverse market movements on a firm's capital and the ability to apply variable rate-of-return targets to business or project evaluation based upon risk considerations.

---

37 EMI, Energy Management Institute, (2011). Energy Risk Management, January 18-19, New York City Area.

- Control: The Earnings-at-Risk framework described could form the basis for implementing effective policies, developing information to help control and reduce earnings volatility, and establishing a limit structure.[38]

11. Conclusion

About 15 years ago, risk management was mainly about the use of swaps and options to hedge interest rates and commodity prices. Back then, risk management was thought of as a pretty much decentralized or compartmentalized activity that could mainly only help the firm.

But the purview of today's risk manager is much broader; it encompasses all aspects of the corporation, including investment and operating decisions as well as financing anything that affects the level and variability of cash flows going forward. It's about ensuring the company's access to capital and its ability to carry out its strategic plan and, in this sense, it is a critical part of the business model.[39]

Enterprise risk management can become a strategic competitive advantage if it is used to identify specific action steps that enhance performance and optimize risk. It can also influence business strategy by identifying potential adjustments related to previously unidentified opportunities and risks.[40]

With the help of this publication and the associated line of products, companies should be in a position to estimate the impact of market risk on corporate earnings and cash flows.

The energy industry has emerged bloodied but unbowed from the global upheaval that marked the end of the past decade. The wild price movements that characterized the energy and commodities markets have flattened out — though they are still subject to extreme volatility, if within narrower ranges. So are the currencies and equities markets, with which they are inextricably linked. And while the brutal cash crunch has eased, the climate remains a hazardous one — where even sovereign states can present a credit risk, with costs rising accordingly. And industry leaders suddenly become a potential liability for all who trade with, supply, insure or finance them — with a knock-on effect throughout the industry.

In summary, today's rapidly changing marketplace requires improved mechanisms for handling diversity and risk management. For firms in industries

38 Lee, A. Y. (1999). CorporateMetrics Technical Document: RiskMetrics Group.

39 Ryu, C. (2008). Enterprise Risk Management and Firm Value, Wien.

40 S&P, (2009). Integrating enterprise risk management analysis into corporate credit rating: Progress Report.

where significant change is taking place, or firms undergoing major internal change, there is most likely a critical need for ERM in the organization.

References

Akdağ, S.A. and Güler, Ö. (2009). Calculation of wind energy potential and economic analysis by using weibull distribution – a case study from Turkey. Part 1: Determination of Weibull Parameters: Energy Sources, Part B: Economics, Planning, and Policy, 4 (1): 1-8.

Anderson, B. (2005). Morgan Stanley roundtable on enterprise risk management and corporate strategy. Journal of Applied Corporate Finance, (Summer).

Beasley, M.S., Richard C. and Dana, H. (2005). Enterprise risk management: an empirical analysis of factors associated with the extent of implementation: Journal of Accounting and Public Policy, 24(6): 521-531.

Brockett, P.L. (2010). Enterprise risk management. Encyclopaedia of Quantitative Risk Analysis and Assessment: John Wiley & Sons.

Ciorciari M. and Blattner P. (2008). Enterprise risk management maturity-level assessment tool: Chicago: ERM Symposium Publications.

COSO, (2004).Enterprise risk management-integrated framework: 22-58.

Çeliktaş, M.S., Sevgili, T. and Kocar, G. (2009). A snapshot of renewable energy research in Turkey: Renewable Energy, 34(1): 1479-1486.

Clewlow, L. and Strickland, C. (2000). Energy derivatives: pricing & risk management.

Deolitte, (2010). Risk intelligence in the energy & resources industry.

Doodian, J.J., Knott J. and Rech J. (2008). ERM and other considerations related to a principles-based approach.

EMI, Energy Management Institute, (2011). Energy risk management. New York City Area.

Energy Business Reports. (2008). Risk management in energy markets.

Fusaro, P.C. (1998). Energy risk management: hedging strategies and instruments for the international energy markets.

Inhaber, H. (2011). Risk of energy production: Second Edition.

James, T. (2008). Energy markets: price risk management and trading: John Wiley and Sons Ltd.

KPMG (2001). Enterprise risk management: an emerging model for building shareholder value. Australia.

Lee, A.Y. (1999). CorporateMetrics Technical Document: RiskMetrics Group.

Matyjewicz, G. and D'Arcangelo, J. R. (2005). ERM-based auditing.

McVey, M. (2007). What is the likelihood and impact of a BA communicating risk?
MEES, Middle East Economic Survey. (2011). Energy & geopolitical risk: MEES Fundamentals.
Miccolis, J., Hiverly, K. and Merkley, B. (2001). Enterprise risk management: trends and emerging practices. Altamonte Springs, FL: Institute of Internal Auditors Research Foundation.
Morgan Stanley Publication, (2006). Corporate risk management: Journal of Applied Corporate Finance, 18 (4).
New York Stock Exchange (NYSE), (2003). Final NYSE corporate governance rules: NYSE, New York.
O'Donnell, E. (2005). Enterprise risk management: A systems-thinking framework for the event identification, International Journal of Accounting Information Systems, 6(3): 177-195.
Olzak, T. (2006), Security Risk Management, http://adventuresinsecurity.com/blog/2006/02/07/security-risk-management.
Raymond, C. (2008). Guiding principles spell success for ERM: National Underwriter: P&C, 112, (3): 30.
RIMS, (2009). Executive Report on Enterprise Risk Management Technology.
Risk Books, (2005). Energy modelling and the management of uncertainty incisive media: 380.
Ryu, C. (2008). Enterprise Risk Management and Firm Value, Wien.
Sander, D. (2010). The risk of enterprise risk management: Sanders Consulting. LLC, Certified Public Accounting Firm. http://www.sas.com/offices/europe/turkey/ (11.10.2010).
S&P (2009). Integrating enterprise risk management analysis into corporate credit rating: Progress Report.
Linsmeier, T.J. and Pearson, N.D. (1996). Risk Measurement: An introduction to value at risk. University of Illinois at Urbana-Champaign.
Topçu, Y.I. and Ülengin, F. (2004). Energy for the future: An integrated decision aid for the case of Turkey Energy. 29, 137-154.
Zhuang, P. and Hou H. (2008). Knowledge visualization of most prolific countries and hot-topics of enterprise risk management. The 4th IEEE International Conference on Wireless Communications, Networking and Mobile Computing.
Wisegeek, (2010). What is energy risk management http://www.wisegeek.com/what-is-energy-risk-management.htm.

Renewable Energy, Turkey and the World

Yusuf Bayraktutan, Sefer Ucak and Selahattin Kaynak

Introduction

The word energy originates from the Ancient Greek expression "energia", meaning "doing or being something". From this point of view, the first modern physics used the word energy for the ability of doing something[1]. Energy has evolved into a compulsory input, not only of industry but also of economic and social life. For this reason, the total energy consumed by human beings is constantly increasing, and environmental values are being severely affected by the increasing consumption of fossil fuels. Reduction in the reserves of fossil fuels, such as, mainly, coal, oil and natural gas, and the environmental problems caused by the energy obtained from these sources generate a need for alternative energy sources.

This study addresses the strategic importance of energy, and the increasing role of alternative energy sources for the world and Turkey.

1. Energy Consumption and Renewable Energy Requirements

Energy is a strategically important input. The countries that own energy sources have an impact on other countries by controlling price and the amount of production. It is a widely accepted observation that there has been an increase in world energy demand, energy prices, and in parallel, the search for alternative energy since the 1960s. Between the years 1980-2007, the consumption of renewable energy sources increased by 8.2% while world energy demand increased by nearly 2.3%[2]. It is expected that most of the world energy demand will be met from developing countries, and the share of non-renewable energy sources in energy consumption will be 82% in 2030. It is also foreseen that oil and natural gas will be the energy sources that are consumed most. The net electric consumption per capita is 2450 Kwh around the world, approximately 7600 in OECD countries, and 1656 in Turkey[3]. Among the OECD countries, electric consumption per capita is lowest in Turkey.

1 Erich Übelacker, **Enerji** (Turkish version by Ali Ulvi Erdoğan), İzmir: TUDEM Y., 2005, p. 4.
2 IEA, **World Energy Outlook 2008**, Paris: IEA P., p. 78.
3 İsmail Yilmaz Aslan, et al., **Enerji Hukuku**, Bursa: Ekin Y., 2007, p. 19.

Technology is a primary requirement for turning underground and surface energy sources into usable energy. Only a small amount of energy sources could be used without technology. Sources of direct use, such as wood, the energy coming from the sun, and thermal heating, don't require advanced technology as they are primary sources. However, some kind of mechanical improvements have enabled human beings to benefit from wind and water energies. Generating energy and electricity from sources such as the wind, sun and nuclear energy is possible merely through intense technological improvements. Using renewable energy sources more efficiently and securely depends on the related technologies. Policy recommendations in this respect could be specified as follows[4]:

i. Research and development activities should be determined in common with the government and the firms in this sector. These activities should contain various shareholders. In this way, it will be possible to decrease the costs in the short and long term.
ii. Technological data for use in strategic research and development planning must contain cost-benefit opportunities, reliability, environmental impact, and progress.
iii. National and international cooperation should be enabled to develop the technical and security standards for new types of energy.
iv. The research and development findings must be shared with the public by means of publications, technical reports and conferences to raise the awareness and usage of renewable energy.

Energy transferring technology, as well as energy producing technology, is one of the factors that affects energy supply. Energy can be transferred from its place of production to the place it is consumed or processed using appropriate technology.

Renewable sources could be used in various ways, such as generating electricity from rivers, heating from solar power, drying agricultural productions, and making use of wind power.

The main problems facing renewable energy sources are the high cost of installation, instability in production, low productivity, and difficulties in storing the generated energy[5].

As seen in Table 1, over time, the production cost of renewable energy sources is expected to decrease, and the annual electricity generating capacity to increase. The cost of generating heat from renewable sources is lower than that of generating electricity, and is expected to decrease further in the future. However, the difference between the costs of generating electrical energy from fossil sources and renewable

4 İbrahim Dinçer, "Renewable Energy and Sustainable Development: A Crucial Review", **Renewable and Sustainable Energy Reviews,** Vol. 4, 2000, pp. 157-175.

5 İsmet Akova, **Yenilenebilir Enerji Kaynaklari**. Ankara: Nobel Y., 2008, p. 19.

energy sources is converging. In the long term, technological innovations and incentives could shift this difference in favour of renewable energies.

Table 1: Cost of Electricity and Heat Generation from Renewable Energy Sources

	Electricity				Heat		
Source	**Production Cost**	**Possible cost in the future**	**Investment Cost**	**Annual capacity increase**	**Production cost**	**Possible cost in the future**	**Investment Cost**
	(cent/kwh)	(cent/kwh)	($/kwh)	(%)	(cent/kwh)	(cent/kwh)	($/kwh)
Biomass	5-15	4-10	900-3000	3	1-5	1-5	250-750
Geothermal	2-10	1-8	800-3000	4	0.5-5	0.5-5	200-2000
Wind	5-13	3-10	1100-1700	30	5-13	3-10	1100-1700
Solar Battery	25-125	5-25	5000-10000	30	3-20	2-10	500-1700
Solar Heat	12-18	4-10	3000-4000	5	–	–	–
Tide	8-15	8-15	1700-2500	0	–	–	–

Source: Ingvar B. Fridleifsson, "Geothermal Energy for the Benefit of the People", Renewable and Sustainable Energy Reviews, Vol. 5, 2001, pp. 299-312.

As seen in Table 2, the generation cost of renewable energies such as biomass, geothermal and wind are more economical than other energy sources. The generation costs of renewable sources, such as solar and hydroelectric are higher than non-renewable sources. However, the cost of renewable sources could be decreased through more investment, incentives, developing technology and wider usage.

Table 2: A Comparison between the Costs of Renewable Energy Sources and Other Energy Sources

	Generation Costs (1 Kwh Energy / Cent)	
Source	**Minimum**	**Maximum**
Coal	4.5	7.0
Natural Gas	4.4	5.0
Geothermal	4.3	6.8
Biomass	4.2	7.9
Wind	4.7	7.2
Solar Energy	6.0	7.8
Nuclear	5.3	9.3
Hydro	5.2	18.9

Source: A. Demirbaş, "Turkey's Renewable Energy Policy", Energy Sources, Vol. 27, No. 3, 2006, p. 662.

The renewable energy sources all around the world could be used in various areas, such as electricity, heating and fuel oil production. Yet it is especially used for electricity generation (approximately 1.350 Gw).

Table 3: World's Renewable Energy Sources (2007)

Source	**Capacity of Electricity Generation (Gw)**
Hydro energy (Total) Big hydroelectric plants Small hydroelectric plants	1250 1010 240
Geothermal energy	9.5
Solar energy	0.4
Wind energy	95
Sea wave energy	0.3
Biomass energy	45
Hot water and heating (GWth)	
Biomass heat	235
Solar collectors	105
Geothermal heating	33
Fuel Oil Production (billion litres/year)	
Ethanol production	46
Biodiesel production	8

Source: Renewable Energy Policy Network For The 21st Century (REN 21), Renewables, Global Status Report, 2007, p. 37.

As seen in Table 3, about 373 Gw of biomass, solar collectors and geothermal energy were used to generate hot water and heating in 2007, while 54 billion litres of biodiesel and ethanol were used as fuel oil.

Most of the world's energy requirement is provided from fossil fuels, with just 18% coming from renewable sources. Again, just 18.4% of all electricity is generated from renewable sources and the rest from fossil sources.

Table 4: World's Energy Consumption and Electricity Generation in Terms of Sources (2007)

Kind of Energy	Share (%)	Electricity generation (%)
Fossil oils	79	67
Renewable sources (total)	18	18.4
Traditional biomass	13	-
Big hydroelectric plants	3	15
Hot water and heating	1.3	-
Wind and solar	0.8	3
Biofuels	0.3	0.4
Nuclear energy	3	14.6
Total	**100**	**100**

Source: Renewable Energy Policy Network for the 21st Century (REN 21), Renewables, Global Status Report, 2007, p. 9.

In 2006, the energy generated from renewable sources amounted to more than 10.8 billion tons equivalent of petrol (TEP); which was 11% of primary energy, and 26% of total energy production. In Turkey, the most important share of renewable energy generation belongs to biomass and hydroelectric. The shares of wind and solar power are very low, but they are expected to increase in the future. Turkey has important renewable energy sources such as hydro, wind, solar, geothermal and biomass.

Table 5: Turkey's Potential Renewable Energy Sources

Renewable Sources	Technique (Mw)	Economic (Mw)
Hydro energy	36.000	13.395
Geothermal energy (heat-electricity)	31.500	5.500 (4.000+1.500)
Solar energy	56.000	-
Wind power (electricity/Set Installed capacity power)	2.500	837
Sea wave energy (electricity)	9.000	-
Biomass energy (fuel, Mtep/Year)	-	16.9
Total	**134.000**	**19.640**

Source: Enerji ve Tabii Kaynaklar Bakanliği, "Yenilenebilir Enerji", www.enerji.gov.tr/ 05.01.2010.

2. Renewable Energy Sources

2.1 Hydroelectric Energy

Hydroelectric plants could turn the potential energy in water into electrical power. However, building dams and hydroelectric plants is expensive and takes a long time. Watering cultivated areas, preventing floods, fishing in dam lakes and water sports are other positive aspects of hydraulic sources. In terms of negative influence, they damage the historical and natural environment of the area where they are built, harm the natural structure of lakes and rivers, and lead to climate change.

Table 6: World's Installed Hydroelectric Power (2009)

Continent	**Hydroelectric Installed Power (%)**
Africa	3
America	36
West Asia and Middle East	9
Europe	29
East Asia and Oceania	23
World Total	**100**

Source: International Hydropower Association, "2010 Activity Report", London: IHA P., p. 5.

Across all continents, Africa has the least installed capacity, and as clearly shown in Table 6, America has the largest installed power capacity. Around the world, the USA (86%), Japan (78%), Norway (68%), Canada (56%) and Turkey (21%) have developed their technical hydroelectric potential. The International Energy Agency (IEA) expects the share of hydroelectric and other renewable energy sources in world energy consumption to have increased by 53% in 2020, compared to the level of 2009. The European Union has adopted an activity plan within the EU Strategies in order to increase renewable energy in Europe, by doubling its share in gross energy consumption (from 6% to 12%), and reaching 22% by 2020.

Table 7: Hydroelectricity (HE) Potential in the World and Turkey

	Gross HE Potential (Gwh/year)	**Technical HE Potential (Gwh/year)**	**Economic HE Potential (Gwh/year)**
World	40 150 000	14 060 000	8 905 000
Europe	3 150 000	1 225 000	800 000
Turkey	433 000	216 000	130 000

Source: DSİ, "Enerji", http://www.dsi.gov.tr/hizmet/enerji.htm/ 09.01.2008.

The maximum ability to generate hydroelectricity that is calculated by assuming all sources are fully used shows that country's hydroelectric potential. However, as it is impossible to use all this potential with present technology, the technical (achievable) potential must be evaluated. The share of technical potential that could be improved within present and expected local economic situations is called economic hydroelectric potential.

Turkey's theoretical hydroelectric potential is 15% of the world's theoretical potential while its economic potential is 16% of Europe's economic potential. Turkey's theoretical hydroelectric potential is estimated to be 433 billion Kwh, its technical potential is 215 billion Kwh, and its technical and economic potential is approximately 130 billion Kwh. The EU's supporting policies and tax reductions for green energy will push up the economic potential. In 2009, there were 150 hydroelectric plants in Turkey. These plants have an established power of 13395 Mw and annual approximate generation capacity of 48100 Gwh, equivalent to 37% of annual total generation. 40 hydroelectric plants with installed power of 3497 Mw and generation power of 112709 Gwh, equivalent to 9% of total generation, are still under construction. A further 526 hydroelectric plants will be built in Turkey in the future in order to use the remaining potential of 70563 Gwh, producing a total of 716 hydroelectric plants with 36697 Mw total installed power. [6] Among the various energy sources, hydroelectric plants are preferred because they offer several advantages and are low risk. Hydroelectric plants are domestic sources that are environmentally friendly, clean, renewable, effective, durable, home-based, and the resulting energy doesn't have unstable prices, high operating costs or require fuel consumption.

Due to management policy, faults, maintenance/repairs, and drought, etc., 73% of the total capacity could be used. 68% of all thermal and 94% of all hydroelectric capacity is utilized. 25% of Turkey's energy is generated by hydroelectric plants, which could qualify as renewable sources, and 75% from thermal sources. Hydroelectric energy is expected to increase 37% to 130 billion Kwh by 2023.[7]

2.2 Wind Energy

Wind energy is created through the replacement of air masses with different temperatures. 1-2% of the energy arriving on earth from the sun turns into wind energy. Wind turbines transform the renewable air flow into electric energy.

6 DSİ, "Enerji", http://www.dsi.gov.tr/hizmet/enerji.htm/ 05.01.2009.

7 DSİ, "Enerji", http://www.dsi.gov.tr/hizmet/enerji.htm/ 05.01.2009.

Wind turbines play an important role in preventing climate change as they do not emit gas. Unlike traditional power plants, wind energy is a low-cost source in terms of energy security, stable prices in the long run, and economical usage. It also decreases dependency on other countries, and could be used anytime. However, there are certain problems associated with wind turbines, such as the large area they cover, the noise they make, and other problems related to the generated electricity. Wind power could be used for generating electricity, charging batteries, storing water, transportation, water pumping, grain milling and cooling.[8]

Table 8: Installed Wind Power Capacity around the World (2007)

Regions	**Megawatt (Mw)**	**Rate (%)**
Africa and Middle East	528	0.5
Asia	16,091	17.1
Europe (Total) EU (27)	57,136 56,535-483	61 60.5-0.5
Latin America and Caribbean	535	0.5
North America	18,664	19.8
Pacific Region	1,158	1.2
World Total	**94,112**	**100**

Source: EIEI, "Rüzgâr Enerjisi Çalişmalari", http://www.eie.gov.tr/ 09.01.2008.

As the table shows, more than half of the installed wind power around the world is in Europe, mainly because of incentives granted for renewable energy, especially in EU countries, and for investment and geographical reasons. North America is home to approximately 20% of all wind power capacity, while Africa and the Middle East seem to have the least wind power capacity, at 0.5%, due to their rich petrol reserves and unsuitable geographical conditions.

Table 9: The Top 10 Countries in the World in terms of Installed Wind Power Capacities (2007)

	Total Capacity (Mw)	**Market Share (%)**
Germany	22,247	23.6
USA	16,818	17.9
Spain	15,145	16.1
India	8,000	8.5

8 Martin Kaltschmitt, W. Streicher, A. Wiese (Ed.), **Renewable Energy**, Berlin: Springer, 2007, p. 49.

Table 9: The Top 10 Countries in the World in terms of Installed Wind Power Capacities (2007) – continued

	Total Capacity (Mw)	**Market Share (%)**
China	6,050	6.4
Denmark	3,125	3.3
Italy	2,726	2.9
France	2,454	2.6
England	2,389	2.5
Portugal	2,150	2.3
Total of First 10 Countries	**81,104**	**86.2**
Rest of the World	**13,008**	**13.8**
World Total	***94,112***	***100***

Source: European Wind Energy Association, Global Statistic, GWEC, 2007, p. 3.

Total world wind output is calculated at 53 Twh/year, and total installed capacity was 94,112 Mw in 2009, of which nearly 24 was in Germany. The other two important countries are the USA and Spain, which together account for 34% of total world wind power. As Table 9 shows, ten countries represent 86% of the world's total capacity. The investment needed to supply 1.245 Gw of wind power by is 692 billion Euro. During this period, the generation costs are expected to decrease from 3.79 €-cents/kwh to 2.45 €-cents/kwh. It is possible to make economically sound investments in regions with total potential of at least 48.00 Mw at least, and an annual average of more than 7.5 m/s.[9]

The Turkey Wind Energy Potential Atlas (REPA), which was drawn up in 2007, indicates that there is at least 50,000 Mw of wind power potential in regions where the annual wind power is 8.5 m/s or more, and there is at least 48,000 Mw of wind energy potential in regions where the wind power is over 7.0 m/s. Since 2004, the level of installed wind power has improved from its initial 18 Mw. Turkey's installed wind power reached 400 Mw in 2008. Since the Renewable Energy Law came into force, 80 new wind projects comprising 2,887 Mw of installed power have been licensed. Among these projects, plants totalling 1,000 Mw of installed power are under construction.[10]

9 EIEI, "Rüzgar Enerjisi", http://www.eie.gov.tr/turkce/YEK/ruzgar/25.12.2009.

10 TÜREB, "Rüzgar Ölçümleri", http://www.ruzgarenerjisibirligi.org.tr/bilgibank-ruzgar-olcum.htm/ 05.01.2008.

In a study conducted in 2002, Turkey's technical wind potential was calculated to be 90,000, and its economic potential to be 10,000 Mw.[11] As shown in Table 10, by the end of 2008, the installed wind power provided by wind plants that are operating and are expected to operate was 840 Mw.

Table 10: Wind Power Plants in Operation or Planned to be put into Use in Turkey

Place	**Company**	**Date of Starting Operation**	**Installed Power (MW)**	**Power Tribune Used**	**Wind turbine Installed Power**	**Number**
İzmir-Çeşme	Alize A.Ş.	1998	1.50	Enercon	600 kW	3
İzmir-Çeşme	Güçbirliği A.Ş.	1998	7.20	Vestas	600 kW	12
Çanakkale-Bozcaada	Bores A.Ş.	2000	10.20	Enercon	600 kW	17
İstanbul-Hadimköy	Sunjüt A.Ş.	2003	1.20	Enercon	600 kW	2
Balikesir-Bandirma	Bares A.Ş.	I/2006	30.00	GE	1.5 MW	20
İstanbul-Silivri	Ertürk A.Ş.	II/2006	0.85	Vestas	850 kW	1
İzmir-Çeşme	Mare A.Ş.	I/2007	39.20	Enercon	800 kW	49
Çanakkale-İntepe	Anemon A.Ş.	I/2007	30.40	Enercon	800 kW	38
Manisa-Akhisar	Deniz A.Ş.	I/2007	10.80	Vestas	1.8 MW	6
Çanakkale-Gelibolu	Doğal A.Ş.	II/2007	15.20	Enercon	880 kW	18
Operating Installed Power			**146.25**			
Manisa-Sayalar	Doğal A.Ş.	II/2007	30.40	Enercon	800 kW	38
Hatay-Samandağ	Deniz A.Ş.	II/2007	30.00	Vestas		
İstanbul-G.paşa	Lodos A.Ş.	I/2008	24.00	Enercon E82	2MW	12
İstanbul-Çatalca	Ertürk A.Ş.	I/2008	60.00	Vestas V90	3 MW	20
Projects under Construction			**144.40**			
Contracted Installed Power (will be put into Use)			**531.66**			
General Total			***837.61***			

Source: EIEI, "Rüzgar Enerjisi Çalişmalari", http://www.eie.gov.tr/ 12.01.2008.

11 Murat Kenisarin, Vedat M.Karsli, and Mehmet Çağlar, "Wind Power Engineering in World and Perspective of Its Development in Turkey", **Renewable and Sustainable Energy Reviews**, Vol. 10-4, 2006, pp. 341-369.

2.3 Geothermal Energy

Geothermal energy could be defined as hot water or steam that is formed by heat that is present underground, and which is always hotter than atmospheric temperature, and which has a higher content of dissolved minerals, various kinds of salt and gases than the normal underground or surface water around it.[12] Geothermal energy is used directly or for electricity generation. Geothermal sources can provide high or low capacity for direct use. Around the world, geothermal energy is more commonly used directly rather than for electricity generation. This is especially true in places that are close to hot water, since transferring hot water is very expensive. The longest hot water network is in Iceland with a length of 63 km. The energy is used directly as follows:[13] hot spring (baths and treatment, 42%), building heating (23%), geothermal heating pumps (12%), greenhouses (19%), fish farms (6%), industry (6%), drying, ice melting and air conditioning (2%).

The problems associated with generating energy from geothermal sources include corrosion, rotting, calcification, and the occurrence of gases that could harm the air, such as carbon dioxide and sulphur.

Table 11: World's Geothermal Energy

	Suitable Capacity for Electricity Generation (Twh)	**Share (%)**
Europe	1830	17
Asia	2970	26
Africa	1220	12
North America	1330	13
South and Middle America	2800	25
Oceania	1050	7
Wold Total	***11200***	***100***

Source: M. H. Dickson, and M. Fanelli, What is Geothermal Energy, Italy: John Wiley&Sons, 2004, p. 57.

Asia and South America account for 51% of the world's total energy generated from geothermal sources. These two continents are followed by Europe, including Turkey, with a share of 7%. When we look at the changes in direct usage of world geothermal energy over time we see that it has reached the equivalent of 30,000 Mw thermal temperatures in 2007. This amount is nearly double what it

---

12 Akova, op.cit., p. 119.

13 Ingvar B. Fridleifsson, "Geothermal Energy For The Benefit of The People", **Renewable And Sustainable Energy Reviews**, Vol. 5, 2001, pp. 291-312.

was five years earlier. 28,000 Mw of energy are generated through direct use of geothermal energy, while 9,100 Mw are currently used for electricity generation. When compared with 2001, this equates to an increase of 14%. Between 1987 and 2007, electricity generation increased by approximately 2.8% annually on average, while direct use increased at a rate of 20% annually.[14]

The Turkish Institute of Mining Technical Research (MTA) states that the number of geothermal sources with over 100 ^{0}C temperature is more than 600. After studying 124 of these, it is estimated that Turkey's thermal potential is 1045 Mw; however this could be doubled easily.[15] In 2005, the amount used as heat was 1230 Mw, of which house heating accounted for 48%, hot springs for treatment and tourism for 40%, and greenhouse heating for the remaining 12%. Nearly 20 Mw of electricity were generated from geothermal sources. This capacity is expected to increase up to 1000 Mw by 2020.[16]

2.4 Solar Energy

Directly or indirectly, the sun is the main source of all energy types, as it is the basic source of life on earth. Solar energy is a clean source, being the energy type which emits the least carbon dioxide into the atmosphere. Solar energy is preferred for the following reasons: It is an infinite source of energy, free (except for the first investment costs), clean, abundant, self-sufficient, achievable in the requested areas because there is no transportation problem, stable in terms of supply and demand, and easy to be used with simple technology.

Although the benefits of solar energy cover a large area and different sectors, the main aim is to use it as much as possible instead of fossil fuels. Solar energy is used in three main areas:[17]

i. The systems used in order to obtain low temperatures with solar energy are as follows: vacuum piped collectors, straight collectors, solar pools, water stiller in pools, heating houses, greenhouses, drying, cooling, solar cookers, water pumping, mechanical power generation.
ii. Generating electricity from solar energy: electricity generation using intensive systems (parabolic ducting collectors, parabolic antenna systems, central receiver systems), solar flues, satellite systems.
iii. Electricity generation using solar batteries.

---

14 IEA, "World Energy Outlook 2007", 2008, p. 593.

15 MTA, "Doğal Kaynaklar ve Ekonomi Bülteni", Sayi: 4, 2007, pp. 22-23.

16 Akova, op.cit., pp. 144-145.

17 Cédric Philibert, The Present and Future Use Of Solar Thermal Energy As a Primary Source of Energy, Paris: IEA Yayini, 2005, pp. 16-20.

The cost of generating electricity from solar energy is high and the first setup cost is about $ 5,000-6,000 per kilowatt (Kw). However, these generation costs decrease over time. These technologies convert only 14-24% of sunlight into energy, indicating low productivity. These are the most important factors preventing the development of solar energy usage.[18]

Table 12 shows that 43% of the world's solar battery capacity is in Europe, and 41% in Asia. Japan is home to 87% of Asia's total solar battery capacity, while 83% of Europe's capacity is in Germany, and 93% of North America's total solar battery capacity is in the USA. Developed countries account for 84% of total power, with the remainder being shared among developing countries.

Solar energy usage has some downsides, as follows[19]:

i. Large areas are required because solar radiation on a specific area is low.
ii. Storage is necessary because solar radiation is not stable, and storage opportunities are limited.
iii. When a lot of energy is needed in winter, solar radiation is very low and absent at nights.
iv. In order to benefit properly from the sunshine, the systems should be clear and without shadow.
v. The initial installation cost is high and not economical.

Table 12: World's Installed Capacity of Solar Batteries (2007)

Place	Installed Capacity (Kwp)	Country	Installed Capacity (Kwp)
Europe	1,705,133	Germany	*1,429,000*
Asia	1,627,311	Japan	*1,421,000*
North America	531,420	U.S.A	*496,000*
South America	27,794	India	*85,000*
Africa	12,288	China	*70,000*
Middle East	1,844	Australia	*60,581*
Oceania	60,581	Other	*404,790*
World total	***3,966,371***	***World total***	***3,966,371***

Source: D. Yogi Goswami, Survey of Energy Resources 2007, London: World Energy Council P., p. 34.

According to a study conducted by the Turkish General Directorate of Electric Power Resources Survey and Development Administration (EIEI) using sun-

18 Akova, op.cit., p. 66.
19 Akova, op.cit., p. 75.

shine duration and shining power statistics that were measured between 1966 and1982 by the General Directorate of State Meteorology (DMI), the sun shines for approximately 2640 hours in Turkey each year (7.2 hours a day), and solar radiation power totals approximately 1311 kwh/m² a year (3.6 kwh/m² a day). The solar energy systems in Turkey are hot water generating systems, most of which are used in Mediterranean and Aegean regions.

Nearly 12 billion m² of solar collectors are installed in Turkey, and annual generation is 750,000 m². A specific amount of this generated energy is exported. Heating energy generated from solar energy totals about 420,000 TEP per year.

Table 13: Turkey's Energy Generation from Solar Collectors.

Years	Solar Energy Generation (Thousand Tep)
1998	210
1999	236
2000	262
2001	290
2004	375
2007	420

Source: EIEI, "Türkiye'de Güneş Enerjisi", http://www.eie.gov.tr/turkce/gunes/tgunes.html 09.01.2009.

Solar batteries could still be used economically in regions that are not close to built-up areas and which are still not connected to the electricity grid. However, they can be used for applications, such as signalling, providing electricity in rural areas, etc., thanks to the possibility of adjusting their capacity. In Turkey, most of the installed solar battery power of 1 Mw is used by the public sector, such as the Ministry of Forestry, forest observation towers, Turkish Telecom, lighthouses and street lighting, EIE, Mugla University, or Aegean University.[20]

2.5 Biomass Energy

Biomass energy could be defined as a type of solar energy that is converted by plants. Biomass energy is produced from the energy that is converted and stored within plants, including fast-growing trees such as sallow, poplar, acacia, vegetation, such as panicum, and agricultural plants, such as sugar cane, industrial cannabis, tobacco, corn, and soya.[21]

20 EIEI, "Türkiye'de Güneş Enerjisi", http://www.eie.gov.tr/turkce/gunes/tgunes.html/ 10.01.2009.

21 Acaroğlu, op.cit., pp. 106-108.

Table 14: World's Biomass Potential (2007)

	Potential (Billion Tones)	%		Potential (Billion Tones)	%
Europe	–	0	Brazil	46	*27*
Asia	62	37	India	24	*15*
North America	24	14	China	18	*9*
South America	58	34	Other	81	*49*
Africa	16	9			
Middle East	–	0			
Oceania	9	6			
World's total	**169**	**100**	**World's total**	**169**	**100**

Source: Ralph P. Overend, Biomass and Bioenergy, World Energy Council Publications, 2007, pp. 67-69.

Asia and Europe rank top with a total share of 71% of the world's biomass potential (Table 14). Brazil, India and China account for 51% of total world potential.

Biomass energy can be categorized in two groups.[22] The first group is classic biomass, such as biogas, comprised of wood, plant, and animal waste. Biogas is a colourless, scentless gas mixture that burns with a blue flame, contains methane, carbon dioxide, hydrogen sulphur and a little nitrogen, and is created by the fermentation of organic waste in an environment without oxygen. The second group is biofuels that are produced from agricultural production and waste oils using various chemical methods, and used in mixtures with benzene or diesel. Biofuels are classified as biodiesel and bioethanol. Biodiesel is the final product obtained as a result of a reaction of frying oil, animal oil or oils of seed plants like soya, coconut, canola with an alcohol such as ethanol by means of a catalyst contributing to the speed of reaction and the quality of production.[23]

Germany accounts for nearly 50% of the world's biodiesel production, followed by French with nearly 15%. European countries have made it compulsory to add some biodiesel to normal diesel fuel, and have decided to permanently increase this rate. Countries that are active in biodiesel production, such as Germany, Italy and France, provide tax exemption to support biodiesel production. Biodiesel production in the USA is very low compared with these countries, and accounts for just 7% of the world's total production. The installed capacity of biodiesel, biogas, and bioethanol in Turkey is 1.5 billion tons, 15 Mw, and 160,000 tons respectively.[24]

22 K. Kaygusuz, and M. F. Türker, "Technical Note, Biomass Energy Potential in Turkey", **Renewable Energy**, Vol. 26, 2002, pp. 661-678.

23 Günnur Koçar, **Biyogaz**, Ankara: Bilim ve Teknik Y., 2006, p. 43.

24 EIEI, Biyoenerji, http://www.eie.gov.tr/biyoenerji/ 11.01.2009.

Table 15: World's Biodiesel Production (2005)

Countries	**Production (Billion Litres)**	***Production (Billion Gallons)***
Germany	1,921	*507*
France	557	*147*
U.S.A	284	*75*
Italy	227	*60*
Czech Republic	136	*36*
Austria	85	*22*
Spain	84	*22*
Denmark	80	*21*
Poland	80	*21*
England	74	*20*
Brasilia	70	*18*
Australia	57	*15*
Sweden	7	*2*
Other Countries	102	*27*
World	**3,762**	**994**

Source: F.O. Licht, World Ethanol and Biofuels Report, Agra Informa Ltd., 2006, p. 54.

The USA produces 40% of the world's bioethanol, followed by Brazil with 37%. Brazil is the most important bioethanol producer in Latin America, and 1/3 of its vehicles run on ethanol obtained from molasses. Countries such as China, Thailand, Malaysia, India and the Philippines plan both to reduce their energy consumption, and to develop their agriculture by using more biodiesel produced from molasses, palm oil and animal oil. Its share among renewable sources is, however, low because of negative factors, such as the air pollution caused by biomass energy, and the devastation of forests.[25]

Biofuel harvest rates shouldn't be higher than biofuel growing rates in order to ensure the durability of the energy obtained from forests and weeds. When it comes to the fertility of biofuels, organic substance clustering and mineralization of the soil should be considered to determine the amount of waste coming from agriculture and forest ecosystems. Some gas emissions occur when biofuels burn. [26]

---

25 İbrahim Yüksel, "Global Warming and Renewable Energy Sources for Sustainable Development in Turkey", **Renewable Energy**, Vol. 33-4, 2008, pp. 802-812.

26 F. Evrendilek and C. Ertekin "Assessing The Potential of Renewable Energy Sources in Turkey", Renewable Energy, Vol. 28, 2003, pp. 2303-2315.

Table 16: World's Bio Ethanol Production Rates (2006)

Countries	Production (Billion Litres)	Countries	*Production (Billion Litres)*
USA	18.4	South Africa	*0.39*
Brazil	17	Spain	*0.46*
China	3.85	England	*0.30*
India	1.9	Thailand	*0.35*
France	0.95	Ukraine	*0.27*
Russia	0.75	Canada	*0.58*
Germany	0.77	Other Countries	*2.1*
Total		**45.9**	

Source: F.O. Licht, World Ethanol and Biofuels Report, Agra Informa Ltd., 2006, p. 56.

2.6 Hydrogen Energy and Fuel Cells

Hydrogen is the most widely available of all elements. More than 90% of the visible substance around the world is made up of hydrogen, which when obtained from water and biomass or fossil sources is considered more an energy transporter than an energy source. Hydrogen is a source of energy which is locally producible, easily and securely transferable to anywhere, and which also has the least energy loss during transfer. It has a broad range of applications, from transportation vehicles to heating. Besides its direct use for fuel-injection engines, hydrogen is also suitable for flameless burning on catalytic surfaces and approximately 33% more fertile than petrol fuels. Transforming hydrogen into energy produces only water vapour, with no harmful gas polluting the environment or increasing the greenhouse effect. It is known that hydrogen is nearly three times more expensive than other fuels and its widespread use depends on technological developments that would lower its cost. Besides, it could be considered as a valid alternative for today to store the redundant electricity generated in daily or seasonal periods as hydrogen.[27]

Developments in the world are in favour of fuel cell technology using hydrogen to enrich. The fuel batteries, used for space research by NASA in the late 1950s, started being used successfully in the industry and service sectors just after the 1990s. Fuel batteries are suitable energy providers for electricity plants as well as being used for mobile devices, such as laptops and mobile phones. Because of their high fertilities and low emissions, they are widely used in transportation sectors.

---

27 Brenda Jonston, Michael C.Mayo and Anshuman Khare, "Hydrogen: The Energy Source for the 21st Century", **Technovation**, Vol. 25-6, 2005, pp. 569-585.

The probably most important qualification of hydrogen is its storage capabilities. There is no suitable method for storing excessive energy, even in this day and age. If the energy generated from hydroelectric plants could be stored, it would be possible to solve the energy problem to some extent. However, the best-known method for storing electricity is none other than acid accumulators. Hydrogen could be stored physically in carbon nano tubes, and chemically as hydride as it is stored purely in tanks as gas or liquid.[28] Hydrogen could be stored in suitably qualified steel tanks as gas or liquid. However, tank weights could cause problems because of the high pressure that occurs when storing gas. The probably cheapest way of storing it is underground in depleted petrol or natural gas reserves, as is done with natural gas. A higher-cost storage solution is the use of mines. Hydrogen has four times the volume of petrol; it should be kept in liquid form to decrease its volume, which requires high pressure and cooling. Liquefied hydrogen could be stored in steel tubes under high pressure. This is the most common method of mesoscale or microscale storage, but very expensive for storing huge amounts since nearly% 25 of the total hydrogen energy is consumed during the liquefaction process. Another solution could be to store liquid hydrogen in low temperature tanks, which is the method that is always used to store the liquid hydrogen used as rocket fuel for space programmes. Studies conducted in the 2000s show that hydrogen could be kept in nano tubes. Solving the weight problem of storing hydrogen in liquid form increases tank volume and cost. Other problems include the losses occurring when hydrogen converts from liquid to gas, and the necessity of a fuel supply. [29]

In energy systems in which hydrogen (which is easy and clean to use in any areas requiring heat and explosion energy) is used as fuel, water and/or water vapour are emitted into the atmosphere. The widespread use of this kind of energy (in mass transportation, for example,) depends on improved automotive technologies based on fuel cells.

Around the world, 500 billion m^3 of hydrogen are produced, stored, transferred and used every year. The total amount of energy used is nearly 79,000 m^3. The largest user is the chemical, and especially the petrochemical, industry.[30]

Fuel batteries are energy converting technologies that are clean, effective and don't harm nature. Electric energy could be generated by chemical reaction without using a vapour tank or turbine. Fuel batteries, obtained from the electrochemical reaction between hydrogen (H_2) and oxygen (O_2), and with total effectiveness of

---

28 Özgür Aslan, "Hidrojen Ekonomisine Doğru", **İstanbul Ticaret Üniversitesi Sosyal Bilimler Dergisi,** Yil: 6, Sayi: 11, 2007, pp. 283-296.

29 EIEI, Hidrojenin Depolanmasi, http://www.eie.gov.tr/turkce/YEK/hidrojen/ 25.02.2010.

30 EIEI, Hidrojen Üretimi, http://www.eie.gov.tr/turkce/YEK/hidrojen/ 25.02.2010.

as much as 80%, are known as durable and electrochemical batteries. Fuel batteries, named according to the kind of electrolyte used within them are better than other forms of energy because they are small in volume, they work effectively and their waste heat can be recovered.[31]

Table 17: World's Hydrogen Using Areas

Usage Areas	**Thousand m^3**
Artificial Fertilizer Industry	25,000 m^3
Herbal Oil (margarine) Production	16,000 m^3
Reserves	1,200 m^3
Petrochemical Industry	30,000 m^3
Hydrogen Animal Oil Production	250 m^3
Gas or Liquid Hydrogen Production	6,000 m^3
Total	**78,450 m^3**

Source: Enerji ve Tabii Kaynaklar Bakanliği, "Hidrojen Enerjisi", http://www.enerji.gov.tr/ 11.01.2008.

2.7 Nuclear Energy

The energy resulting from fission is called "atomic energy" or "nuclear energy". Nuclear reactors are systems that convert nuclear energy into electricity. Basically, nuclear energy, resulting from fission, turns into heat energy within the nuclear fuel and other substances; and this heat energy turns into kinetic energy and then into electricity in a generator system.[32]

Table 18: World's Nuclear Reactors

Country	**Operating**		**Under construction**	
	Number of Units	**Total Mwe**	**Number of Units**	**Total Mwe**
U.S.A	104	100582	1	1165
Germany	17	20470	.	.
Argentina	2	935	1	692
Belgium	7	5824	.	.
Brazil	2	1795	.	.
Bulgaria	2	1906	2	1906

31 EIEI, "Yakit Pilleri", http://www.eie.gov.tr/turkce/YEK/hidrojen/yakit_pilleri.html/20.02.2010.
32 Erich Übelacker, op.cit., pp. 22-23.

Table 18: World's Nuclear Reactors – continued

Country	Operating		Under construction	
	Number of Units	**Total Mwe**	**Number of Units**	**Total Mwe**
Czech Republic	6	3619	.	.
China	11	8572	6	5220
Taiwan	6	4921	2	2600
Armenia	1	376	.	.
Finland	4	2696	1	1600
France	59	63260	1	1600
South Africa	2	1800	.	.
India	17	3782	6	2910
Holland	1	482	.	.
England	19	10222	.	.
Iran	.	.	1	915
Spain	8	7450	.	.
Sweden	10	9014	.	.
Switzerland	5	3220	.	.
Japan	55	47587	1	866
Canada	18	12589	.	.
Korea Republic.	20	17451	3	2880
Lithuania	1	1185	.	.
Hungary	4	1829	.	.
Mexico	2	1360	.	.
Pakistan	2	425	1	300
Romania	2	1300	.	.
Russia	31	21743	6	3639
Slovakia	5	2034	.	.
Slovenia	1	666	.	.
Ukraine	15	13107	2	1900
Total	**439**	**371936**	**34**	**28193**

Source: TAEK, "Nükleer Enerji", http://www.taek.gov.tr/bilgi/sss/durum.html/12.01.2008.

The fuel used at nuclear plants is kept within the plant for 10-20 years. During this period it loses 98% of its activity. Durable radioactive substances, which are the main problem, are vitrified and put into lead, concrete and corrosion-resistant protective coverings, where they are then stored in specific places 1,000 metres beneath the surface. A 1,000 Mw nuclear reactor produces 27 tons ($7m^3$) of waste fuel annually. Fossil fuels, especially coal plants, have a much more negative effect on the environment that can't even be compared with nuclear plants. Nuclear plants are preferable in terms of their environmental effect; the negative effect of nuclear reactors working under normal operating conditions is limited to 50.1-1 of the normal natural radiation level; however, the figure is lower in practice.[33]

In terms of the sustainability of generating electricity, nuclear plants are considered to be more secure than thermal and hydroelectric plants. Around the world, besides the developments in widening the usage of renewable energy sources, projects directed at nuclear energy investments have gained speed. Electricity for commercial use started to be generated in nuclear plants in the 1950s. The total capacity of the 439 nuclear reactors in operation in 32 countries all around the world is 372 Gw. The countries that own most nuclear plants are the USA, France, Japan, and Russia. These four countries own 35% of the world's plants – 149 plants – and generate 60% of the total power produced from nuclear energy. Nuclear power meets 16% of the world's energy needs.[34]

In Turkey, 5,000 Mw of nuclear plant capacity have been planned as of 2015, based on electricity supply and demand projections. Accordingly, laws were introduced to regulate the installation and operation of nuclear power plants and the sale of energy in 2007. The process of installing nuclear power plants is in progress. The license for the first plant of Turkey, which is planned for installation in Mersin-Akkuyu, has been provided, and licensing studies continue for Sinop.[35]

Conclusion

The main principles to be kept in mind when deciding on a country's energy policies are as stated: Energy should be seamless and continuous; energy sources should be trustworthy, energy should be delivered on time; it should be environment friendly and clean, from generation to consumption.

While planning a country's energy policies in the light of these principles,[36] it is necessary to view the correlation among energy, economy and ecology, coor-

33 Atilla Sandikli and Hasret Dikici Bilgin (Ed.), **Sürdürülebilir Kalkinma için Nükleer Enerjinin Önemi,** İstanbul: TASAM Y., 2006, pp. 144-146.

34 Yemane Wolde-Rufael, and Kojo Menyah, "Nuclear Energy Consumption and Economic Growth in Nine Developed Countries", **Energy Economics**, Vol. 32, Issue: 3, 2010, pp. 550-556.

35 Enerji ve Tabii Kaynaklar Bakanliği, "Nükleer Enerji", http://www.enerji.gov.tr/12.01.2009

36 Necdet Pamir, "Enerji Politikalari ve Küresel Gelişmeler", **Stratejik Analiz,** (Aralik) 2005, pp. 68-69.

dinate the planning of sustainable development, create source variety, maintain balances, evaluate geopolitical and geographic factors, and provide energy security.

This sequence of importance will help to avoid problems associated with energy and the various energy sources in different areas of economic activity: energy needs should be met with a variety of energy sources and types; besides evaluating present sources, the use of renewable energy sources should be enabled; domestic production should be increased with the use various technologies; efficiency in energy production and consumption should be increased, and environmental factors should be considered at each step; the restructuring of the energy sector should be based on cleanliness and competition.

In the light of all these evaluations, technological improvements in energy systems and services, taking more advantage of renewable energy sources, preferring clean production technologies when consuming fuels, empowering the correlation between economy and the environment, and coordinating at international level should be the primary components of energy policies.

References

Acaroğlu, Mustafa. Alternatif Enerji Kaynaklari, Ankara: Nobel Y., 2007.

Aslan, İsmail Yilmaz, et.al., Enerji Hukuku, Bursa: Ekin Y., 2007.

Aslan, Özgür, "Hidrojen Ekonomisine Doğru", İstanbul Ticaret Üniversitesi Sosyal Bilimler Dergisi, Yil: 6, Sayi: 11, 2007, pp. 283-296.

Demirbaş, Ayhan, "Turkey's Renewable Energy Policy", Energy Sources, Vol. 27, No. 3, 2006, pp. 657-665.

DSİ (Devlet Su İşleri), "Enerji", http://www.dsi.gov.tr/hizmet/enerji.htm/ 05.01.2009.

Dickson, Mary H., and Mario Fanelli, What is Geothermal Energy, Italy: John Wiley and Sons P. , 2004.

DiNÇER, İbrahim, "Renewable Energy and Sustainable Development: A Crucial Review", Renewable and Sustainable Energy Reviews, Vol. 4, 2000, pp. 157-175.

EİEİ (Elektrik İşleri Etüt İdaresi), "Rüzgâr Enerjisi", http://www.eie.gov.tr/turkce/YEK/ ruzgar/25.12.2009

EİEİ, "Türkiye'de Güneş Enerjisi", http://www.eie.gov.tr/turkce/gunes/tgunes.html/ 10.01.2009.

EİEİ, "Yakit Pilleri", http://www.eie.gov.tr/turkce/YEK/hidrojen/yakit_pilleri.html/ 20.02.2010.

EİEİ, "Biyoenerji", http://www.eie.gov.tr/biyoenerji/ 11.01.2009.

EİEİ, "Hidrojen Üretimi", http://www.eie.gov.tr/turkce/YEK/hidrojen/ 25.02.2010.
ETKB (TC Enerji ve Tabii Kaynaklar Bakanliği), "Nükleer Enerji", http://www.enerji.gov.tr/ 12.01.2009.
ETKB, "Yenilenebilir Enerji", www.enerji.gov.tr/ 05.01.2010.
European Wind Energy Association, Global Statistics, GWEC, 2007.
Evrendilek, F., and C. Ertekin, "Assessing The Potential of Renewable Energy Sources in Turkey" Renewable Energy, Vol. 28, 2003, pp. 2303-2315.
Fridleifsson, Ingvar B., "Geothermal Energy for the Benefit of the People", Renewable and Sustainable Energy Reviews, Vol. 5, 2001, pp. 291-312.
Goswami, D.Yogi, Survey of Energy Resources 2007, London: World Energy Council p. , 2007.
IEA, "World Energy Outlook 2007", Paris: IEA, 2008.
IEA, "World Energy Outlook 2008", Paris: IEA, 2009.
International Hydropower Association, 2010 Activity Report, London: IHA P., 2010.
Jonston, Brenda, Michael C. Mayo, and Anshuman Khare, "Hydrogen: The Energy Source for the 21st Century", Technovation, Vol. 25-6, 2005, pp. 569-585.
Kaltschmitt, Martin W., Streicher, and A. Wiese (Eds), Renewable Energy, Berlin: Springer, 2007.
Kaygusuz, K., and M. F. Türker, "Technical Note, Biomass Energy Potential in Turkey", Renewable Energy, Vol. 26, 2002, pp. 661-678.
Kenisarin, Murat, Vedat M. Karsli, and Mehmet Çağlar, "Wind Power Engineering in World and Perspective of Its Development in Turkey", Renewable and Sustainable Energy Reviews, Vol. 10-4, 2006, pp. 341-369.
Koçar, Günnur, Biyogaz, Ankara: Bilim ve Teknik Y., 2006.
Licht, F.O., World Ethanol and Biofuels Report, Agra Informa Ltd., 2006.
MTA, "Doğal Kaynaklar ve Ekonomi Bülteni", Sayi: 4, 2007, ss. 22-23.
Overend, Ralph P. Biomass and Bioenergy, World Energy Council Publications, 2007.
Pamir, Necdet, "Enerji Politikaları ve Küresel Gelişmeler", Stratejik Analiz, (Aralik) 2005, pp. 68-69.
Philibert, Cédric, The Present and Future Use of Solar Thermal Energy as a Primary Source of Energy, Paris: IEA p. , 2005.
Renewable Energy Policy Network for the 21st Century (REN 21), Renewables, Global Status Report, 2007.
Sandikli, Atilla, and Hasret Dikici Bilgin (Ed.), Sürdürülebilir Kalkinma için Nükleer Enerjinin Önemi, İstanbul: TASAM Y., 2006.
TAEK, "Nükleer Enerji", http://www.taek.gov.tr/bilgi/sss/durum.html/ 12.01.2008.

TÜREB, “Rüzgar Ölçümleri”, http://www.ruzgarenerjisibirligi.org.tr/bilgibank-ruzgar-olcum. htm/ 05.01.2008.

Übelacker, Erich, Enerji (Turkish version by Ali Ulvi Erdoğan), İzmir: TUDEM Y., 2005.

Yemane, Wolde-Rufael, and Kojo Menyah, “Nuclear Energy Consumption and Economic Growth in Nine Developed Countries”, Energy Economics, Vol. 32, Issue: 3, 2010, pp. 550-556.

Yüksel, İbrahim, “Global Warming And Renewable Energy Sources for Sustainable Development in Turkey”, Renewable Energy, Vol. 33-4, 2008, pp. 802-812.

About the Authors

Selahattin Kaynak
Bayburt University Faculty of Economics and Administrative Sciences.
Department of Economics
skaynak@bayburt.edu.tr

İbrahim Arslan
Gaziantep University Faculty of Economics and Administrative Sciences.
Department of Economics
iarslan@gantep.edu.tr

A. Yilmaz Ata
Gaziantep University Faculty of Economics and Administrative Sciences.
Department of Economics
ayata@gantep.edu.tr

Abdulkadir Develi
Yıldırım Beyazıd University Faculty of Political Sciences.
Department of Economics
adeveli@bayburt.edu.tr

Ahmet Ay
Selçuk University Faculty of Economics and Administrative Sciences.
Department of Economics
ahmetay@selcuk.edu.tr

Zeynep Karaçor
Selçuk University Faculty of Economics and Administrative Sciences.
Department of Economics
zkaracor@selcuk.edu.tr

Şaban Çalış
Selçuk University Faculty of Economics and Administrative Sciences. Department of International Relations
scalis@selcuk.edu.tr

Yuri Yegorov
University of Vienna Department of Industry Energy and Environment
yury.egorov@univie.ac.at

Franz Wirl
University of Vienna Department of Industry Energy and Environment
franz.wirl@univie.ac.at

Yeşim Reel
Marmara University Faculty of Economics and Administrative Sciences. Department of Economics
yesimreel@gmail.com

Gökhan Özkan
Yalova University Faculty of Economics and Administrative Sciences. Department of International Relations
gozkan@yalova.edu.tr

Müslüme Narin
Gazi University Faculty of Economics and Administrative Sciences. Department of Economics
muslume@gazi.edu.tr

Yusuf Bayraktutan
Kocaeli University Faculty of Economics and Administrative Sciences. Department of Economics
ybayraktutan@kocaeli.edu.tr

Hakki M. Ay
KAM University Faculty of Economics and Administrative Sciences. Department of Economics
hma@kmu.edu.tr

Nazan Şahbaz
KAM University Faculty of Economics and Administrative Sciences.
Department of Economics
nsahbaz@kmu.edu.tr

Sefer Ucak
Balikesir University Sindirgi Vocational School
sucak@balikesir.edu.tr

Bünyamin Topçu
PhD.
bunyamintopcu@yahoo.com

Şule Güneş
PhD. Candidate
sulegunes@gmail.com

FSC
www.fsc.org
MIX
Papier | Fördert gute Waldnutzung
FSC® C083411

Kolibri 360 GmbH
Ferdinand-Jühlke-Straße 7
99095 Erfurt, Deutschland
produktsicherheit@kolibri360.de